Table of contents

Section 1

Name:...............................   Date:...............................

Find the sum

Complete all the activities (Addition).

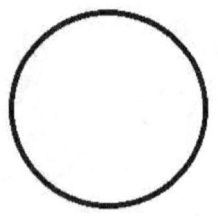

SCORE

| 1. | 600
+ 191 | 2. | 981
+ 644 | 3. | 700
+ 938 | 4. | 207
+ 900 | 5. | 372
+ 367 | 6. | 959
+ 819 | 7. | 489
+ 460 |

| 8. | 186
+ 866 | 9. | 856
+ 542 | 10. | 214
+ 199 | 11. | 463
+ 482 | 12. | 776
+ 535 | 13. | 816
+ 147 | 14. | 497
+ 835 |

| 15. | 885
+ 647 | 16. | 436
+ 127 | 17. | 168
+ 153 | 18. | 407
+ 196 | 19. | 340
+ 494 | 20. | 360
+ 596 | 21. | 630
+ 372 |

| 22. | 324
+ 791 | 23. | 457
+ 437 | 24. | 132
+ 851 | 25. | 478
+ 851 | 26. | 371
+ 831 | 27. | 742
+ 138 | 28. | 163
+ 218 |

| 29. | 414
+ 546 | 30. | 827
+ 652 | 31. | 542
+ 371 | 32. | 115
+ 724 | 33. | 930
+ 932 | 34. | 751
+ 437 | 35. | 153
+ 989 |

| 36. | 626
+ 462 | 37. | 282
+ 616 | 38. | 568
+ 191 | 39. | 258
+ 984 | 40. | 179
+ 479 | 41. | 115
+ 645 | 42. | 457
+ 135 |

Name:................................   Date:................................

Find the sum

Complete all the activities (Addition).

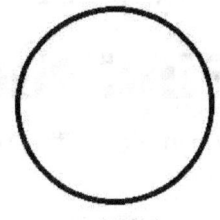

SCORE

1.	600	2.	981	3.	700	4.	207	5.	372	6.	959	7.	489
	+ 191		+ 644		+ 938		+ 900		+ 367		+ 819		+ 460
	791		1 625		1 638		1 107		739		1 778		949

8.	186	9.	856	10.	214	11.	463	12.	776	13.	816	14.	497
	+ 866		+ 542		+ 199		+ 482		+ 535		+ 147		+ 835
	1 052		1 398		413		945		1 311		963		1 332

15.	885	16.	436	17.	168	18.	407	19.	340	20.	360	21.	630
	+ 647		+ 127		+ 153		+ 196		+ 494		+ 596		+ 372
	1 532		563		321		603		834		956		1 002

22.	324	23.	457	24.	132	25.	478	26.	371	27.	742	28.	163
	+ 791		+ 437		+ 851		+ 851		+ 831		+ 138		+ 218
	1 115		894		983		1 329		1 202		880		381

29.	414	30.	827	31.	542	32.	115	33.	930	34.	751	35.	153
	+ 546		+ 652		+ 371		+ 724		+ 932		+ 437		+ 989
	960		1 479		913		839		1 862		1 188		1 142

36.	626	37.	282	38.	568	39.	258	40.	179	41.	115	42.	457
	+ 462		+ 616		+ 191		+ 984		+ 479		+ 645		+ 135
	1 088		898		759		1 242		658		760		592

Name:................................. Date:.................................

Find the sum

Complete all the activities (Addition).

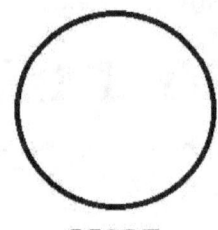

SCORE

1. 665 + 773	2. 891 + 843	3. 907 + 819	4. 941 + 865	5. 321 + 960	6. 691 + 548	7. 326 + 581
8. 759 + 249	9. 688 + 255	10. 883 + 833	11. 650 + 553	12. 600 + 808	13. 582 + 922	14. 561 + 276
15. 758 + 990	16. 530 + 944	17. 318 + 525	18. 157 + 965	19. 114 + 283	20. 195 + 534	21. 209 + 742
22. 408 + 375	23. 732 + 821	24. 185 + 781	25. 695 + 797	26. 254 + 819	27. 310 + 383	28. 816 + 212
29. 272 + 183	30. 362 + 984	31. 357 + 765	32. 207 + 330	33. 530 + 746	34. 330 + 800	35. 524 + 794
36. 235 + 505	37. 159 + 449	38. 723 + 907	39. 245 + 326	40. 176 + 760	41. 838 + 779	42. 580 + 960

Find the sum

Complete all the activities (Addition).

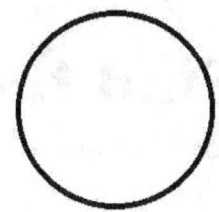

SCORE

1.	665	2.	891	3.	907	4.
	+ 773		+ 843		+ 819	
	1 438		1 734		1 726	

1.   665   2.   891   3.   907   4.   941   5.   321   6.   691   7.   326
   + 773   + 843   + 819   + 865   + 960   + 548   + 581
   1 438   1 734   1 726   1 806   1 281   1 239   907

8.   759   9.   688   10.   883   11.   650   12.   600   13.   582   14.   561
   + 249   + 255   + 833   + 553   + 808   + 922   + 276
   1 008   943   1 716   1 203   1 408   1 504   837

15.   758   16.   530   17.   318   18.   157   19.   114   20.   195   21.   209
   + 990   + 944   + 525   + 965   + 283   + 534   + 742
   1 748   1 474   843   1 122   397   729   951

22.   408   23.   732   24.   185   25.   695   26.   254   27.   310   28.   816
   + 375   + 821   + 781   + 797   + 819   + 383   + 212
     783   1 553   966   1 492   1 073   693   1 028

29.   272   30.   362   31.   357   32.   207   33.   530   34.   330   35.   524
   + 183   + 984   + 765   + 330   + 746   + 800   + 794
     455   1 346   1 122   537   1 276   1 130   1 318

36.   235   37.   159   38.   723   39.   245   40.   176   41.   838   42.   580
   + 505   + 449   + 907   + 326   + 760   + 779   + 960
     740   608   1 630   571   936   1 617   1 540

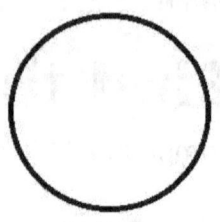

Find the sum

Complete all the activities (Addition).

| 1. | 835
+ 710 | 2. | 734
+ 934 | 3. | 708
+ 944 | 4. | 792
+ 930 | 5. | 920
+ 211 | 6. | 276
+ 443 | 7. | 635
+ 178 |

| 8. | 844
+ 985 | 9. | 590
+ 384 | 10. | 357
+ 900 | 11. | 662
+ 748 | 12. | 760
+ 859 | 13. | 659
+ 909 | 14. | 966
+ 946 |

| 15. | 107
+ 601 | 16. | 246
+ 577 | 17. | 285
+ 246 | 18. | 494
+ 819 | 19. | 112
+ 396 | 20. | 192
+ 579 | 21. | 102
+ 304 |

| 22. | 584
+ 948 | 23. | 273
+ 380 | 24. | 372
+ 167 | 25. | 481
+ 190 | 26. | 269
+ 554 | 27. | 522
+ 679 | 28. | 398
+ 187 |

| 29. | 949
+ 874 | 30. | 375
+ 798 | 31. | 977
+ 846 | 32. | 923
+ 419 | 33. | 964
+ 698 | 34. | 258
+ 489 | 35. | 683
+ 125 |

| 36. | 972
+ 295 | 37. | 292
+ 777 | 38. | 898
+ 458 | 39. | 264
+ 811 | 40. | 687
+ 379 | 41. | 637
+ 900 | 42. | 565
+ 916 |

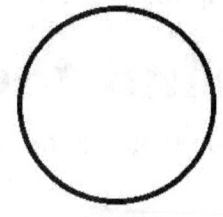

Find the sum

Complete all the activities (Addition).

| 1. | 835
+ 710
1 545 | 2. | 734
+ 934
1 668 | 3. | 708
+ 944
1 652 | 4. | 792
+ 930
1 722 | 5. | 920
+ 211
1 131 | 6. | 276
+ 443
719 | 7. | 635
+ 178
813 |

| 8. | 844
+ 985
1 829 | 9. | 590
+ 384
974 | 10. | 357
+ 900
1 257 | 11. | 662
+ 748
1 410 | 12. | 760
+ 859
1 619 | 13. | 659
+ 909
1 568 | 14. | 966
+ 946
1 912 |

| 15. | 107
+ 601
708 | 16. | 246
+ 577
823 | 17. | 285
+ 246
531 | 18. | 494
+ 819
1 313 | 19. | 112
+ 396
508 | 20. | 192
+ 579
771 | 21. | 102
+ 304
406 |

| 22. | 584
+ 948
1 532 | 23. | 273
+ 380
653 | 24. | 372
+ 167
539 | 25. | 481
+ 190
671 | 26. | 269
+ 554
823 | 27. | 522
+ 679
1 201 | 28. | 398
+ 187
585 |

| 29. | 949
+ 874
1 823 | 30. | 375
+ 798
1 173 | 31. | 977
+ 846
1 823 | 32. | 923
+ 419
1 342 | 33. | 964
+ 698
1 662 | 34. | 258
+ 489
747 | 35. | 683
+ 125
808 |

| 36. | 972
+ 295
1 267 | 37. | 292
+ 777
1 069 | 38. | 898
+ 458
1 356 | 39. | 264
+ 811
1 075 | 40. | 687
+ 379
1 066 | 41. | 637
+ 900
1 537 | 42. | 565
+ 916
1 481 |

Name:............................... Date:...............................

Find the sum

Complete all the activities (Addition).

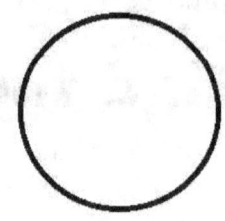

SCORE

| 1. | 432 + 691 | 2. | 155 + 605 | 3. | 847 + 406 | 4. | 868 + 238 | 5. | 121 + 393 | 6. | 601 + 255 | 7. | 379 + 266 |

1. 432
  + 691
  ----

2. 155
  + 605
  ----

3. 847
  + 406
  ----

4. 868
  + 238
  ----

5. 121
  + 393
  ----

6. 601
  + 255
  ----

7. 379
  + 266
  ----

8. 983
  + 574
  ----

9. 222
  + 882
  ----

10. 482
  + 456
  ----

11. 217
  + 974
  ----

12. 362
  + 744
  ----

13. 779
  + 939
  ----

14. 923
  + 238
  ----

15. 287
  + 995
  ----

16. 261
  + 177
  ----

17. 381
  + 108
  ----

18. 986
  + 373
  ----

19. 643
  + 652
  ----

20. 641
  + 377
  ----

21. 256
  + 835
  ----

22. 274
  + 898
  ----

23. 277
  + 743
  ----

24. 179
  + 997
  ----

25. 827
  + 213
  ----

26. 257
  + 754
  ----

27. 517
  + 209
  ----

28. 799
  + 457
  ----

29. 429
  + 410
  ----

30. 825
  + 907
  ----

31. 482
  + 730
  ----

32. 950
  + 125
  ----

33. 232
  + 523
  ----

34. 873
  + 372
  ----

35. 279
  + 710
  ----

36. 929
  + 568
  ----

37. 341
  + 205
  ----

38. 300
  + 381
  ----

39. 800
  + 463
  ----

40. 567
  + 136
  ----

41. 860
  + 925
  ----

42. 410
  + 174

Find the sum

Complete all the activities (Addition).

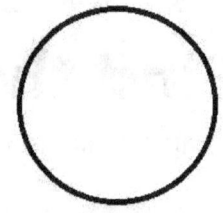

SCORE

1. 432 + 691 1 123	2. 155 + 605 760	3. 847 + 406 1 253	4. 868 + 238 1 106	5. 121 + 393 514	6. 601 + 255 856	7. 379 + 266 645
8. 983 + 574 1 557	9. 222 + 882 1 104	10. 482 + 456 938	11. 217 + 974 1 191	12. 362 + 744 1 106	13. 779 + 939 1 718	14. 923 + 238 1 161
15. 287 + 995 1 282	16. 261 + 177 438	17. 381 + 108 489	18. 986 + 373 1 359	19. 643 + 652 1 295	20. 641 + 377 1 018	21. 256 + 835 1 091
22. 274 + 898 1 172	23. 277 + 743 1 020	24. 179 + 997 1 176	25. 827 + 213 1 040	26. 257 + 754 1 011	27. 517 + 209 726	28. 799 + 457 1 256
29. 429 + 410 839	30. 825 + 907 1 732	31. 482 + 730 1 212	32. 950 + 125 1 075	33. 232 + 523 755	34. 873 + 372 1 245	35. 279 + 710 989
36. 929 + 568 1 497	37. 341 + 205 546	38. 300 + 381 681	39. 800 + 463 1 263	40. 567 + 136 703	41. 860 + 925 1 785	42. 410 + 174 584

Find the sum

Complete all the activities (Addition).

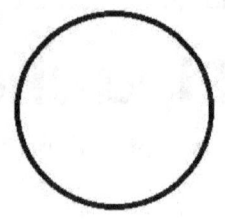

SCORE

1. 837 + 507	2. 659 + 272	3. 308 + 628	4. 943 + 333	5. 826 + 734	6. 894 + 524	7. 884 + 248
8. 766 + 752	9. 869 + 806	10. 818 + 218	11. 283 + 688	12. 513 + 412	13. 907 + 518	14. 656 + 966
15. 843 + 789	16. 146 + 557	17. 420 + 876	18. 287 + 724	19. 864 + 701	20. 164 + 649	21. 766 + 697
22. 741 + 606	23. 310 + 775	24. 496 + 652	25. 695 + 814	26. 266 + 154	27. 207 + 292	28. 890 + 249
29. 993 + 346	30. 364 + 789	31. 971 + 976	32. 489 + 138	33. 687 + 437	34. 204 + 108	35. 960 + 143
36. 888 + 268	37. 742 + 269	38. 451 + 585	39. 946 + 906	40. 202 + 232	41. 122 + 104	42. 370 + 129

Find the sum

Complete all the activities (Addition).

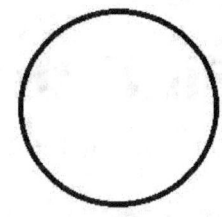

1. 837 + 507 1 344	2. 659 + 272 931	3. 308 + 628 936	4. 943 + 333 1 276	5. 826 + 734 1 560	6. 894 + 524 1 418	7. 884 + 248 1 132
8. 766 + 752 1 518	9. 869 + 806 1 675	10. 818 + 218 1 036	11. 283 + 688 971	12. 513 + 412 925	13. 907 + 518 1 425	14. 656 + 966 1 622
15. 843 + 789 1 632	16. 146 + 557 703	17. 420 + 876 1 296	18. 287 + 724 1 011	19. 864 + 701 1 565	20. 164 + 649 813	21. 766 + 697 1 463
22. 741 + 606 1 347	23. 310 + 775 1 085	24. 496 + 652 1 148	25. 695 + 814 1 509	26. 266 + 154 420	27. 207 + 292 499	28. 890 + 249 1 139
29. 993 + 346 1 339	30. 364 + 789 1 153	31. 971 + 976 1 947	32. 489 + 138 627	33. 687 + 437 1 124	34. 204 + 108 312	35. 960 + 143 1 103
36. 888 + 268 1 156	37. 742 + 269 1 011	38. 451 + 585 1 036	39. 946 + 906 1 852	40. 202 + 232 434	41. 122 + 104 226	42. 370 + 129 499

Name:..............................   Date:..............................

Find the sum

Complete all the activities (Addition).

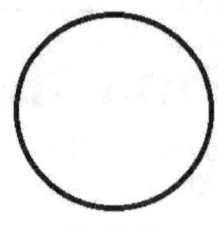

SCORE

1.   622
+ 991
----

2.   450
+ 228
----

3.   985
+ 138
----

4.   730
+ 617
----

5.   437
+ 374
----

6.   677
+ 789
----

7.   307
+ 703
----

8.   918
+ 872
----

9.   671
+ 487
----

10.   691
+ 653
----

11.   925
+ 997
----

12.   412
+ 937
----

13.   840
+ 405
----

14.   302
+ 187
----

15.   244
+ 232
----

16.   767
+ 967
----

17.   706
+ 460
----

18.   515
+ 696
----

19.   986
+ 975
----

20.   122
+ 922
----

21.   412
+ 370
----

22.   831
+ 467
----

23.   591
+ 661
----

24.   378
+ 343
----

25.   987
+ 329
----

26.   561
+ 634
----

27.   413
+ 717
----

28.   203
+ 624
----

29.   679
+ 512
----

30.   387
+ 680
----

31.   834
+ 642
----

32.   473
+ 305
----

33.   607
+ 709
----

34.   183
+ 846
----

35.   289
+ 322
----

36.   885
+ 694
----

37.   465
+ 767
----

38.   341
+ 655
----

39.   266
+ 943
----

40.   447
+ 279
----

41.   127
+ 959
----

42.   199
+ 914

Find the sum

Complete all the activities (Addition).

Name:................................   Date:................................

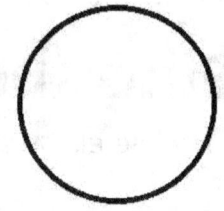

SCORE

1. 622 + 991 1 613	2. 450 + 228 678	3. 985 + 138 1 123	4. 730 + 617 1 347	5. 437 + 374 811	6. 677 + 789 1 466	7. 307 + 703 1 010

8. 918   9. 671   10. 691   11. 925   12. 412   13. 840   14. 302
   + 872   + 487   + 653   + 997   + 937   + 405   + 187
   1 790   1 158   1 344   1 922   1 349   1 245   489

15. 244   16. 767   17. 706   18. 515   19. 986   20. 122   21. 412
    + 232   + 967   + 460   + 696   + 975   + 922   + 370
     476   1 734   1 166   1 211   1 961   1 044   782

22. 831   23. 591   24. 378   25. 987   26. 561   27. 413   28. 203
    + 467   + 661   + 343   + 329   + 634   + 717   + 624
    1 298   1 252   721   1 316   1 195   1 130   827

29. 679   30. 387   31. 834   32. 473   33. 607   34. 183   35. 289
    + 512   + 680   + 642   + 305   + 709   + 846   + 322
    1 191   1 067   1 476   778   1 316   1 029   611

36. 885   37. 465   38. 341   39. 266   40. 447   41. 127   42. 199
    + 694   + 767   + 655   + 943   + 279   + 959   + 914
    1 579   1 232   996   1 209   726   1 086   1 113

Name:................................. Date:.............................

Find the sum

Complete all the activities (Addition).

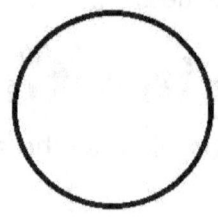

SCORE

| 1. | 670
+ 233 | 2. | 654
+ 988 | 3. | 357
+ 889 | 4. | 116
+ 574 | 5. | 689
+ 441 | 6. | 188
+ 769 | 7. | 896
+ 626 |

| 8. | 860
+ 881 | 9. | 578
+ 531 | 10. | 154
+ 765 | 11. | 443
+ 865 | 12. | 269
+ 645 | 13. | 476
+ 547 | 14. | 458
+ 735 |

| 15. | 165
+ 208 | 16. | 653
+ 875 | 17. | 543
+ 558 | 18. | 633
+ 474 | 19. | 155
+ 193 | 20. | 589
+ 398 | 21. | 306
+ 411 |

| 22. | 106
+ 124 | 23. | 679
+ 571 | 24. | 757
+ 639 | 25. | 183
+ 565 | 26. | 700
+ 595 | 27. | 905
+ 433 | 28. | 570
+ 804 |

| 29. | 963
+ 110 | 30. | 613
+ 809 | 31. | 532
+ 380 | 32. | 509
+ 581 | 33. | 787
+ 312 | 34. | 325
+ 438 | 35. | 992
+ 486 |

| 36. | 826
+ 485 | 37. | 941
+ 754 | 38. | 559
+ 133 | 39. | 860
+ 245 | 40. | 965
+ 904 | 41. | 789
+ 817 | 42. | 652
+ 599 |

Find the sum

Complete all the activities (Addition).

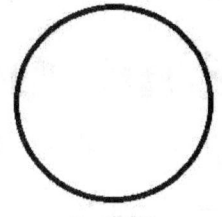

SCORE

1. 670 + 233 903	2. 654 + 988 1 642	3. 357 + 889 1 246	4. 116 + 574 690	5. 689 + 441 1 130	6. 188 + 769 957	7. 896 + 626 1 522
8. 860 + 881 1 741	9. 578 + 531 1 109	10. 154 + 765 919	11. 443 + 865 1 308	12. 269 + 645 914	13. 476 + 547 1 023	14. 458 + 735 1 193
15. 165 + 208 373	16. 653 + 875 1 528	17. 543 + 558 1 101	18. 633 + 474 1 107	19. 155 + 193 348	20. 589 + 398 987	21. 306 + 411 717
22. 106 + 124 230	23. 679 + 571 1 250	24. 757 + 639 1 396	25. 183 + 565 748	26. 700 + 595 1 295	27. 905 + 433 1 338	28. 570 + 804 1 374
29. 963 + 110 1 073	30. 613 + 809 1 422	31. 532 + 380 912	32. 509 + 581 1 090	33. 787 + 312 1 099	34. 325 + 438 763	35. 992 + 486 1 478
36. 826 + 485 1 311	37. 941 + 754 1 695	38. 559 + 133 692	39. 860 + 245 1 105	40. 965 + 904 1 869	41. 789 + 817 1 606	42. 652 + 599 1 251

Name:...................................... Date:......................................

Find the sum

Complete all the activities (Addition).

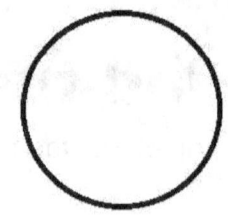

SCORE

1. 507 + 158	2. 969 + 210	3. 987 + 828	4. 510 + 145	5. 580 + 779	6. 866 + 925	7. 901 + 524
8. 335 + 467	9. 346 + 419	10. 647 + 696	11. 738 + 315	12. 653 + 251	13. 873 + 538	14. 943 + 818
15. 144 + 708	16. 120 + 713	17. 727 + 908	18. 174 + 729	19. 343 + 215	20. 450 + 264	21. 975 + 781
22. 904 + 262	23. 888 + 634	24. 231 + 436	25. 909 + 598	26. 658 + 470	27. 271 + 954	28. 860 + 292
29. 504 + 818	30. 336 + 549	31. 884 + 691	32. 101 + 871	33. 759 + 252	34. 297 + 991	35. 449 + 605
36. 224 + 230	37. 968 + 483	38. 720 + 956	39. 410 + 490	40. 362 + 339	41. 407 + 607	42. 908 + 545

Find the sum

Complete all the activities (Addition).

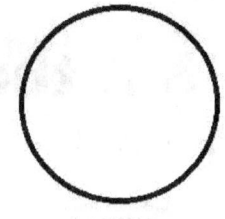

SCORE

| 1. | 507
+ 158
665 | 2. | 969
+ 210
1 179 | 3. | 987
+ 828
1 815 | 4. | 510
+ 145
655 | 5. | 580
+ 779
1 359 | 6. | 866
+ 925
1 791 | 7. | 901
+ 524
1 425 |

| 8. | 335
+ 467
802 | 9. | 346
+ 419
765 | 10. | 647
+ 696
1 343 | 11. | 738
+ 315
1 053 | 12. | 653
+ 251
904 | 13. | 873
+ 538
1 411 | 14. | 943
+ 818
1 761 |

| 15. | 144
+ 708
852 | 16. | 120
+ 713
833 | 17. | 727
+ 908
1 635 | 18. | 174
+ 729
903 | 19. | 343
+ 215
558 | 20. | 450
+ 264
714 | 21. | 975
+ 781
1 756 |

| 22. | 904
+ 262
1 166 | 23. | 888
+ 634
1 522 | 24. | 231
+ 436
667 | 25. | 909
+ 598
1 507 | 26. | 658
+ 470
1 128 | 27. | 271
+ 954
1 225 | 28. | 860
+ 292
1 152 |

| 29. | 504
+ 818
1 322 | 30. | 336
+ 549
885 | 31. | 884
+ 691
1 575 | 32. | 101
+ 871
972 | 33. | 759
+ 252
1 011 | 34. | 297
+ 991
1 288 | 35. | 449
+ 605
1 054 |

| 36. | 224
+ 230
454 | 37. | 968
+ 483
1 451 | 38. | 720
+ 956
1 676 | 39. | 410
+ 490
900 | 40. | 362
+ 339
701 | 41. | 407
+ 607
1 014 | 42. | 908
+ 545
1 453 |

Find the sum

Complete all the activities (Addition).

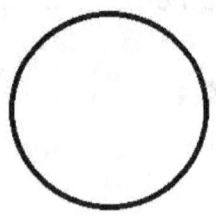

SCORE

1.	150 + 649	2.	543 + 772	3.	148 + 881	4.	325 + 523	5.	878 + 900	6.	379 + 384	7.	781 + 133

8.	503 + 736	9.	436 + 239	10.	986 + 404	11.	474 + 527	12.	501 + 787	13.	334 + 903	14.	881 + 114

15.	556 + 996	16.	243 + 377	17.	782 + 554	18.	821 + 210	19.	708 + 525	20.	943 + 567	21.	380 + 700

22.	351 + 167	23.	204 + 436	24.	275 + 377	25.	987 + 625	26.	902 + 268	27.	620 + 604	28.	172 + 622

29.	247 + 755	30.	538 + 244	31.	999 + 311	32.	966 + 881	33.	758 + 101	34.	601 + 326	35.	311 + 393

36.	127 + 910	37.	294 + 519	38.	124 + 773	39.	276 + 434	40.	682 + 898	41.	728 + 604	42.	126 + 930

Find the sum

Name:................................ Date:................................

Complete all the activities (Addition).

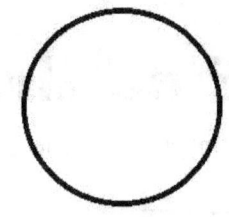

SCORE

1. 150
    + 649
    799

2. 543
    + 772
    1 315

3. 148
    + 881
    1 029

4. 325
    + 523
    848

5. 878
    + 900
    1 778

6. 379
    + 384
    763

7. 781
    + 133
    914

8. 503
    + 736
    1 239

9. 436
    + 239
    675

10. 986
    + 404
    1 390

11. 474
    + 527
    1 001

12. 501
    + 787
    1 288

13. 334
    + 903
    1 237

14. 881
    + 114
    995

15. 556
    + 996
    1 552

16. 243
    + 377
    620

17. 782
    + 554
    1 336

18. 821
    + 210
    1 031

19. 708
    + 525
    1 233

20. 943
    + 567
    1 510

21. 380
    + 700
    1 080

22. 351
    + 167
    518

23. 204
    + 436
    640

24. 275
    + 377
    652

25. 987
    + 625
    1 612

26. 902
    + 268
    1 170

27. 620
    + 604
    1 224

28. 172
    + 622
    794

29. 247
    + 755
    1 002

30. 538
    + 244
    782

31. 999
    + 311
    1 310

32. 966
    + 881
    1 847

33. 758
    + 101
    859

34. 601
    + 326
    927

35. 311
    + 393
    704

36. 127
    + 910
    1 037

37. 294
    + 519
    813

38. 124
    + 773
    897

39. 276
    + 434
    710

40. 682
    + 898
    1 580

41. 728
    + 604
    1 332

42. 126
    + 930
    1 056

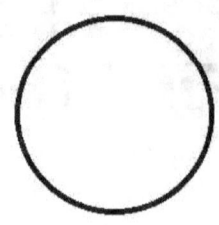

Find the sum

Complete all the activities (Addition).

1.	194 + 974	2.	756 + 279	3.	483 + 299	4.	962 + 952	5.	986 + 799	6.	567 + 168	7.	733 + 469

8.	408 + 203	9.	834 + 948	10.	165 + 694	11.	598 + 991	12.	972 + 201	13.	579 + 398	14.	446 + 812

15.	173 + 639	16.	515 + 494	17.	286 + 136	18.	787 + 783	19.	344 + 423	20.	843 + 113	21.	430 + 174

22.	923 + 494	23.	951 + 955	24.	140 + 806	25.	126 + 579	26.	164 + 779	27.	539 + 431	28.	432 + 202

29.	101 + 655	30.	698 + 185	31.	852 + 249	32.	880 + 447	33.	384 + 173	34.	882 + 697	35.	681 + 624

36.	858 + 511	37.	205 + 153	38.	364 + 525	39.	167 + 579	40.	250 + 623	41.	357 + 639	42.	787 + 592

Find the sum

Complete all the activities (Addition).

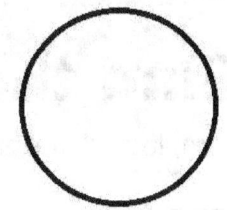

SCORE

Name:................................. Date:.................................

1. 194
    + 974
    1 168

2. 756
    + 279
    1 035

3. 483
    + 299
    782

4. 962
    + 952
    1 914

5. 986
    + 799
    1 785

6. 567
    + 168
    735

7. 733
    + 469
    1 202

8. 408
    + 203
    611

9. 834
    + 948
    1 782

10. 165
    + 694
    859

11. 598
    + 991
    1 589

12. 972
    + 201
    1 173

13. 579
    + 398
    977

14. 446
    + 812
    1 258

15. 173
    + 639
    812

16. 515
    + 494
    1 009

17. 286
    + 136
    422

18. 787
    + 783
    1 570

19. 344
    + 423
    767

20. 843
    + 113
    956

21. 430
    + 174
    604

22. 923
    + 494
    1 417

23. 951
    + 955
    1 906

24. 140
    + 806
    946

25. 126
    + 579
    705

26. 164
    + 779
    943

27. 539
    + 431
    970

28. 432
    + 202
    634

29. 101
    + 655
    756

30. 698
    + 185
    883

31. 852
    + 249
    1 101

32. 880
    + 447
    1 327

33. 384
    + 173
    557

34. 882
    + 697
    1 579

35. 681
    + 624
    1 305

36. 858
    + 511
    1 369

37. 205
    + 153
    358

38. 364
    + 525
    889

39. 167
    + 579
    746

40. 250
    + 623
    873

41. 357
    + 639
    996

42. 787
    + 592
    1 379

Name:......................................   Date:...............................

Find the sum

Complete all the activities (Addition).

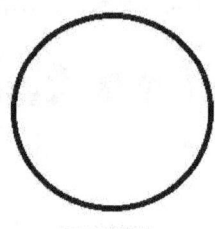

SCORE

1.	139 + 518	2.	458 + 550	3.	190 + 349	4.	476 + 327	5.	103 + 876	6.	989 + 320	7.	393 + 933

8.	498 + 816	9.	621 + 251	10.	264 + 831	11.	980 + 704	12.	587 + 308	13.	553 + 174	14.	971 + 278

15.	547 + 207	16.	222 + 508	17.	440 + 231	18.	467 + 301	19.	804 + 440	20.	647 + 266	21.	202 + 618

22.	529 + 923	23.	401 + 933	24.	952 + 253	25.	278 + 624	26.	675 + 894	27.	179 + 914	28.	925 + 434

29.	930 + 250	30.	996 + 116	31.	377 + 298	32.	779 + 895	33.	754 + 580	34.	879 + 262	35.	565 + 297

36.	112 + 281	37.	603 + 997	38.	728 + 462	39.	275 + 664	40.	906 + 619	41.	221 + 818	42.	100 + 922

Find the sum

Complete all the activities (Addition).

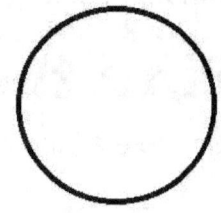

SCORE

1. 139	2. 458	3. 190	4. 476	5. 103	6. 989	7. 393
+ 518	+ 550	+ 349	+ 327	+ 876	+ 320	+ 933
657	1 008	539	803	979	1 309	1 326
8. 498	9. 621	10. 264	11. 980	12. 587	13. 553	14. 971
+ 816	+ 251	+ 831	+ 704	+ 308	+ 174	+ 278
1 314	872	1 095	1 684	895	727	1 249
15. 547	16. 222	17. 440	18. 467	19. 804	20. 647	21. 202
+ 207	+ 508	+ 231	+ 301	+ 440	+ 266	+ 618
754	730	671	768	1 244	913	820
22. 529	23. 401	24. 952	25. 278	26. 675	27. 179	28. 925
+ 923	+ 933	+ 253	+ 624	+ 894	+ 914	+ 434
1 452	1 334	1 205	902	1 569	1 093	1 359
29. 930	30. 996	31. 377	32. 779	33. 754	34. 879	35. 565
+ 250	+ 116	+ 298	+ 895	+ 580	+ 262	+ 297
1 180	1 112	675	1 674	1 334	1 141	862
36. 112	37. 603	38. 728	39. 275	40. 906	41. 221	42. 100
+ 281	+ 997	+ 462	+ 664	+ 619	+ 818	+ 922
393	1 600	1 190	939	1 525	1 039	1 022

Find the sum

Complete all the activities (Addition).

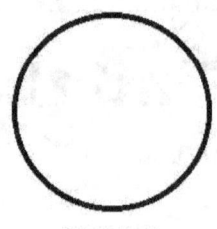

SCORE

| 1. 531
+ 327 | 2. 429
+ 525 | 3. 306
+ 134 | 4. 370
+ 654 | 5. 443
+ 631 | 6. 889
+ 904 | 7. 838
+ 634 |

| 8. 268
+ 820 | 9. 155
+ 790 | 10. 687
+ 980 | 11. 612
+ 494 | 12. 391
+ 182 | 13. 286
+ 562 | 14. 259
+ 366 |

| 15. 343
+ 477 | 16. 894
+ 558 | 17. 297
+ 857 | 18. 464
+ 528 | 19. 251
+ 453 | 20. 955
+ 532 | 21. 926
+ 457 |

| 22. 837
+ 972 | 23. 208
+ 824 | 24. 624
+ 640 | 25. 495
+ 186 | 26. 564
+ 429 | 27. 231
+ 120 | 28. 409
+ 137 |

| 29. 934
+ 348 | 30. 339
+ 743 | 31. 874
+ 104 | 32. 411
+ 965 | 33. 735
+ 431 | 34. 706
+ 980 | 35. 269
+ 740 |

| 36. 669
+ 700 | 37. 835
+ 255 | 38. 153
+ 256 | 39. 656
+ 419 | 40. 972
+ 461 | 41. 938
+ 617 | 42. 870
+ 258 |

Find the sum

Complete all the activities (Addition).

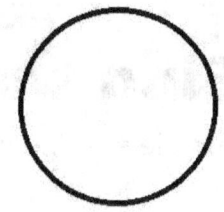

SCORE

| 1. | 531
+ 327
858 | 2. | 429
+ 525
954 | 3. | 306
+ 134
440 | 4. | 370
+ 654
1 024 | 5. | 443
+ 631
1 074 | 6. | 889
+ 904
1 793 | 7. | 838
+ 634
1 472 |

| 8. | 268
+ 820
1 088 | 9. | 155
+ 790
945 | 10. | 687
+ 980
1 667 | 11. | 612
+ 494
1 106 | 12. | 391
+ 182
573 | 13. | 286
+ 562
848 | 14. | 259
+ 366
625 |

| 15. | 343
+ 477
820 | 16. | 894
+ 558
1 452 | 17. | 297
+ 857
1 154 | 18. | 464
+ 528
992 | 19. | 251
+ 453
704 | 20. | 955
+ 532
1 487 | 21. | 926
+ 457
1 383 |

| 22. | 837
+ 972
1 809 | 23. | 208
+ 824
1 032 | 24. | 624
+ 640
1 264 | 25. | 495
+ 186
681 | 26. | 564
+ 429
993 | 27. | 231
+ 120
351 | 28. | 409
+ 137
546 |

| 29. | 934
+ 348
1 282 | 30. | 339
+ 743
1 082 | 31. | 874
+ 104
978 | 32. | 411
+ 965
1 376 | 33. | 735
+ 431
1 166 | 34. | 706
+ 980
1 686 | 35. | 269
+ 740
1 009 |

| 36. | 669
+ 700
1 369 | 37. | 835
+ 255
1 090 | 38. | 153
+ 256
409 | 39. | 656
+ 419
1 075 | 40. | 972
+ 461
1 433 | 41. | 938
+ 617
1 555 | 42. | 870
+ 258
1 128 |

Find the sum

Complete all the activities (Addition).

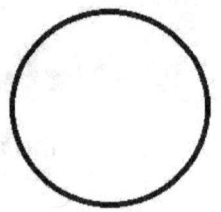

SCORE

1. 912 + 565	2. 427 + 278	3. 547 + 427	4. 592 + 902	5. 214 + 312	6. 657 + 923	7. 146 + 871
8. 336 + 777	9. 218 + 643	10. 471 + 755	11. 516 + 863	12. 594 + 975	13. 965 + 254	14. 480 + 616
15. 374 + 878	16. 804 + 357	17. 233 + 253	18. 153 + 270	19. 104 + 529	20. 533 + 841	21. 248 + 766
22. 587 + 623	23. 597 + 229	24. 210 + 783	25. 182 + 296	26. 343 + 395	27. 246 + 215	28. 741 + 149
29. 970 + 451	30. 311 + 394	31. 946 + 976	32. 628 + 335	33. 507 + 400	34. 790 + 892	35. 818 + 166
36. 773 + 688	37. 210 + 629	38. 607 + 749	39. 376 + 970	40. 478 + 836	41. 144 + 696	42. 533 + 179

Find the sum

Complete all the activities (Addition).

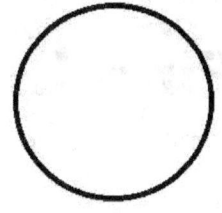

1. 912 + 565 1 477	2. 427 + 278 705	3. 547 + 427 974	4. 592 + 902 1 494	5. 214 + 312 526	6. 657 + 923 1 580	7. 146 + 871 1 017
8. 336 + 777 1 113	9. 218 + 643 861	10. 471 + 755 1 226	11. 516 + 863 1 379	12. 594 + 975 1 569	13. 965 + 254 1 219	14. 480 + 616 1 096
15. 374 + 878 1 252	16. 804 + 357 1 161	17. 233 + 253 486	18. 153 + 270 423	19. 104 + 529 633	20. 533 + 841 1 374	21. 248 + 766 1 014
22. 587 + 623 1 210	23. 597 + 229 826	24. 210 + 783 993	25. 182 + 296 478	26. 343 + 395 738	27. 246 + 215 461	28. 741 + 149 890
29. 970 + 451 1 421	30. 311 + 394 705	31. 946 + 976 1 922	32. 628 + 335 963	33. 507 + 400 907	34. 790 + 892 1 682	35. 818 + 166 984
36. 773 + 688 1 461	37. 210 + 629 839	38. 607 + 749 1 356	39. 376 + 970 1 346	40. 478 + 836 1 314	41. 144 + 696 840	42. 533 + 179 712

Name:................................   Date:................................

Find the sum

Complete all the activities (Addition).

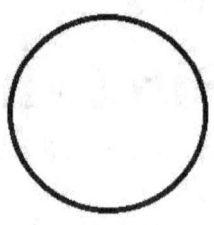

SCORE

| 1. 634
+ 509 | 2. 210
+ 118 | 3. 511
+ 105 | 4. 299
+ 652 | 5. 317
+ 239 | 6. 894
+ 271 | 7. 238
+ 351 |

| 8. 215
+ 296 | 9. 438
+ 317 | 10. 965
+ 116 | 11. 522
+ 654 | 12. 285
+ 617 | 13. 607
+ 991 | 14. 772
+ 100 |

| 15. 972
+ 474 | 16. 350
+ 351 | 17. 542
+ 549 | 18. 700
+ 997 | 19. 290
+ 772 | 20. 480
+ 813 | 21. 867
+ 221 |

| 22. 471
+ 970 | 23. 734
+ 436 | 24. 612
+ 125 | 25. 855
+ 282 | 26. 498
+ 825 | 27. 730
+ 517 | 28. 656
+ 979 |

| 29. 323
+ 592 | 30. 509
+ 519 | 31. 332
+ 299 | 32. 680
+ 842 | 33. 888
+ 643 | 34. 728
+ 687 | 35. 650
+ 217 |

| 36. 106
+ 665 | 37. 836
+ 252 | 38. 248
+ 401 | 39. 283
+ 315 | 40. 882
+ 271 | 41. 479
+ 747 | 42. 244
+ 790 |

Find the sum

Complete all the activities (Addition).

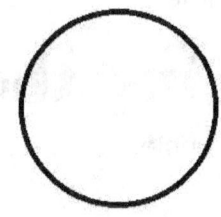

SCORE

1. 634 + 509 = 1 143	2. 210 + 118 = 328	3. 511 + 105 = 616	4. 299 + 652 = 951	5. 317 + 239 = 556	6. 894 + 271 = 1 165	7. 238 + 351 = 589

1.   634   2.   210   3.   511   4.   299   5.   317   6.   894   7.   238
+ 509   + 118   + 105   + 652   + 239   + 271   + 351
1 143   328   616   951   556   1 165   589

8.   215   9.   438   10. 965   11. 522   12. 285   13. 607   14. 772
+ 296   + 317   + 116   + 654   + 617   + 991   + 100
511   755   1 081   1 176   902   1 598   872

15. 972   16. 350   17. 542   18. 700   19. 290   20. 480   21. 867
+ 474   + 351   + 549   + 997   + 772   + 813   + 221
1 446   701   1 091   1 697   1 062   1 293   1 088

22. 471   23. 734   24. 612   25. 855   26. 498   27. 730   28. 656
+ 970   + 436   + 125   + 282   + 825   + 517   + 979
1 441   1 170   737   1 137   1 323   1 247   1 635

29. 323   30. 509   31. 332   32. 680   33. 888   34. 728   35. 650
+ 592   + 519   + 299   + 842   + 643   + 687   + 217
915   1 028   631   1 522   1 531   1 415   867

36. 106   37. 836   38. 248   39. 283   40. 882   41. 479   42. 244
+ 665   + 252   + 401   + 315   + 271   + 747   + 790
771   1 088   649   598   1 153   1 226   1 034

© KingSchool Edition

Find the sum

Complete all the activities (Addition).

Name:..............................   Date:.............................

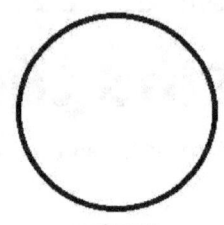

SCORE

1.	976	2.	620	3.	892	4.	774	5.	363	6.	581	7.	540
	+ 524		+ 125		+ 720		+ 697		+ 960		+ 668		+ 628

8.	182	9.	707	10.	473	11.	276	12.	988	13.	369	14.	796
	+ 191		+ 923		+ 737		+ 831		+ 578		+ 355		+ 109

15.	710	16.	140	17.	154	18.	865	19.	123	20.	794	21.	442
	+ 338		+ 584		+ 889		+ 464		+ 703		+ 929		+ 408

22.	402	23.	669	24.	860	25.	233	26.	563	27.	374	28.	614
	+ 875		+ 579		+ 654		+ 791		+ 555		+ 506		+ 245

29.	535	30.	754	31.	923	32.	986	33.	920	34.	614	35.	762
	+ 142		+ 637		+ 765		+ 125		+ 996		+ 651		+ 716

36.	168	37.	229	38.	395	39.	529	40.	908	41.	494	42.	874
	+ 104		+ 543		+ 866		+ 213		+ 790		+ 242		+ 918

Find the sum

Complete all the activities (Addition).

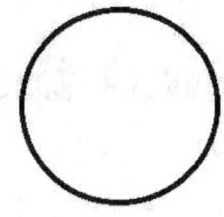
Name:..   Date:..

1.	976	2.	620	3.	892	4.	774	5.	363	6.	581	7.	540
	+ 524		+ 125		+ 720		+ 697		+ 960		+ 668		+ 628
	1 500		745		1 612		1 471		1 323		1 249		1 168

8.	182	9.	707	10.	473	11.	276	12.	988	13.	369	14.	796
	+ 191		+ 923		+ 737		+ 831		+ 578		+ 355		+ 109
	373		1 630		1 210		1 107		1 566		724		905

15.	710	16.	140	17.	154	18.	865	19.	123	20.	794	21.	442
	+ 338		+ 584		+ 889		+ 464		+ 703		+ 929		+ 408
	1 048		724		1 043		1 329		826		1 723		850

22.	402	23.	669	24.	860	25.	233	26.	563	27.	374	28.	614
	+ 875		+ 579		+ 654		+ 791		+ 555		+ 506		+ 245
	1 277		1 248		1 514		1 024		1 118		880		859

29.	535	30.	754	31.	923	32.	986	33.	920	34.	614	35.	762
	+ 142		+ 637		+ 765		+ 125		+ 996		+ 651		+ 716
	677		1 391		1 688		1 111		1 916		1 265		1 478

36.	168	37.	229	38.	395	39.	529	40.	908	41.	494	42.	874
	+ 104		+ 543		+ 866		+ 213		+ 790		+ 242		+ 918
	272		772		1 261		742		1 698		736		1 792

Find the sum

Complete all the activities (Addition).

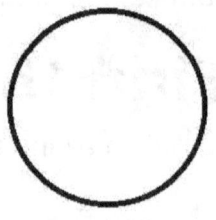

SCORE

1.	774 + 584	2.	310 + 647	3.	374 + 720	4.	928 + 980	5.	947 + 256	6.	106 + 676	7.	421 + 378

8.	955 + 672	9.	964 + 309	10.	891 + 516	11.	723 + 559	12.	426 + 830	13.	211 + 747	14.	714 + 422

15.	233 + 964	16.	243 + 781	17.	890 + 169	18.	350 + 436	19.	786 + 966	20.	376 + 142	21.	787 + 187

22.	156 + 972	23.	537 + 718	24.	546 + 252	25.	257 + 336	26.	825 + 285	27.	350 + 459	28.	295 + 315

29.	258 + 480	30.	536 + 627	31.	106 + 606	32.	847 + 983	33.	113 + 324	34.	786 + 352	35.	128 + 619

36.	885 + 997	37.	932 + 849	38.	728 + 636	39.	503 + 453	40.	393 + 492	41.	301 + 563	42.	265 + 896

Find the sum

Complete all the activities (Addition).

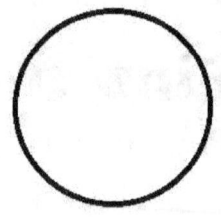

1. 774 + 584 1 358	2. 310 + 647 957	3. 374 + 720 1 094	4. 928 + 980 1 908	5. 947 + 256 1 203	6. 106 + 676 782	7. 421 + 378 799
8. 955 + 672 1 627	9. 964 + 309 1 273	10. 891 + 516 1 407	11. 723 + 559 1 282	12. 426 + 830 1 256	13. 211 + 747 958	14. 714 + 422 1 136
15. 233 + 964 1 197	16. 243 + 781 1 024	17. 890 + 169 1 059	18. 350 + 436 786	19. 786 + 966 1 752	20. 376 + 142 518	21. 787 + 187 974
22. 156 + 972 1 128	23. 537 + 718 1 255	24. 546 + 252 798	25. 257 + 336 593	26. 825 + 285 1 110	27. 350 + 459 809	28. 295 + 315 610
29. 258 + 480 738	30. 536 + 627 1 163	31. 106 + 606 712	32. 847 + 983 1 830	33. 113 + 324 437	34. 786 + 352 1 138	35. 128 + 619 747
36. 885 + 997 1 882	37. 932 + 849 1 781	38. 728 + 636 1 364	39. 503 + 453 956	40. 393 + 492 885	41. 301 + 563 864	42. 265 + 896 1 161

Find the sum

Complete all the activities (Addition).

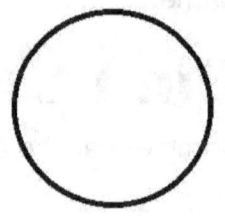

1. 239 + 957	2. 532 + 531	3. 873 + 134	4. 278 + 406	5. 530 + 748	6. 136 + 281	7. 339 + 172
8. 959 + 313	9. 931 + 294	10. 224 + 637	11. 450 + 826	12. 352 + 500	13. 730 + 658	14. 812 + 804
15. 187 + 144	16. 426 + 542	17. 441 + 950	18. 843 + 155	19. 791 + 355	20. 478 + 993	21. 441 + 578
22. 568 + 284	23. 813 + 547	24. 396 + 430	25. 493 + 180	26. 560 + 107	27. 439 + 627	28. 546 + 412
29. 705 + 132	30. 800 + 315	31. 321 + 472	32. 319 + 486	33. 422 + 593	34. 838 + 388	35. 328 + 904
36. 520 + 676	37. 815 + 230	38. 296 + 871	39. 359 + 168	40. 686 + 284	41. 362 + 482	42. 411 + 107

Find the sum

Complete all the activities (Addition).

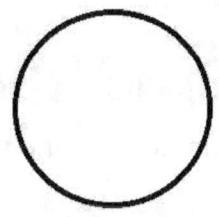

SCORE

| 1. | 239
+ 957
1 196 | 2. | 532
+ 531
1 063 | 3. | 873
+ 134
1 007 | 4. | 278
+ 406
684 | 5. | 530
+ 748
1 278 | 6. | 136
+ 281
417 | 7. | 339
+ 172
511 |

| 8. | 959
+ 313
1 272 | 9. | 931
+ 294
1 225 | 10. | 224
+ 637
861 | 11. | 450
+ 826
1 276 | 12. | 352
+ 500
852 | 13. | 730
+ 658
1 388 | 14. | 812
+ 804
1 616 |

| 15. | 187
+ 144
331 | 16. | 426
+ 542
968 | 17. | 441
+ 950
1 391 | 18. | 843
+ 155
998 | 19. | 791
+ 355
1 146 | 20. | 478
+ 993
1 471 | 21. | 441
+ 578
1 019 |

| 22. | 568
+ 284
852 | 23. | 813
+ 547
1 360 | 24. | 396
+ 430
826 | 25. | 493
+ 180
673 | 26. | 560
+ 107
667 | 27. | 439
+ 627
1 066 | 28. | 546
+ 412
958 |

| 29. | 705
+ 132
837 | 30. | 800
+ 315
1 115 | 31. | 321
+ 472
793 | 32. | 319
+ 486
805 | 33. | 422
+ 593
1 015 | 34. | 838
+ 388
1 226 | 35. | 328
+ 904
1 232 |

| 36. | 520
+ 676
1 196 | 37. | 815
+ 230
1 045 | 38. | 296
+ 871
1 167 | 39. | 359
+ 168
527 | 40. | 686
+ 284
970 | 41. | 362
+ 482
844 | 42. | 411
+ 107
518 |

Find the sum

Complete all the activities (Addition).

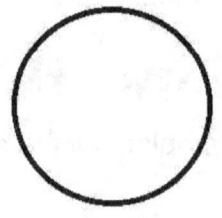

SCORE

1. 583 + 419	2. 203 + 814	3. 504 + 531	4. 448 + 760	5. 554 + 263	6. 771 + 968	7. 359 + 182
8. 860 + 320	9. 707 + 108	10. 202 + 589	11. 659 + 883	12. 291 + 309	13. 430 + 674	14. 153 + 638
15. 444 + 616	16. 952 + 839	17. 161 + 313	18. 902 + 103	19. 490 + 125	20. 883 + 796	21. 555 + 760
22. 222 + 652	23. 458 + 327	24. 612 + 276	25. 306 + 467	26. 575 + 354	27. 380 + 822	28. 447 + 699
29. 228 + 893	30. 483 + 829	31. 478 + 810	32. 416 + 937	33. 310 + 932	34. 907 + 146	35. 279 + 857
36. 829 + 683	37. 224 + 405	38. 133 + 898	39. 761 + 638	40. 851 + 203	41. 161 + 876	42. 431 + 977

Find the sum

Complete all the activities (Addition).

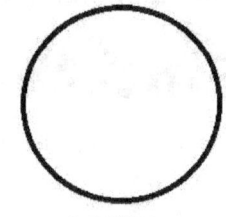

SCORE

1. 583 + 419 1 002	2. 203 + 814 1 017	3. 504 + 531 1 035	4. 448 + 760 1 208	5. 554 + 263 817	6. 771 + 968 1 739	7. 359 + 182 541
8. 860 + 320 1 180	9. 707 + 108 815	10. 202 + 589 791	11. 659 + 883 1 542	12. 291 + 309 600	13. 430 + 674 1 104	14. 153 + 638 791
15. 444 + 616 1 060	16. 952 + 839 1 791	17. 161 + 313 474	18. 902 + 103 1 005	19. 490 + 125 615	20. 883 + 796 1 679	21. 555 + 760 1 315
22. 222 + 652 874	23. 458 + 327 785	24. 612 + 276 888	25. 306 + 467 773	26. 575 + 354 929	27. 380 + 822 1 202	28. 447 + 699 1 146
29. 228 + 893 1 121	30. 483 + 829 1 312	31. 478 + 810 1 288	32. 416 + 937 1 353	33. 310 + 932 1 242	34. 907 + 146 1 053	35. 279 + 857 1 136
36. 829 + 683 1 512	37. 224 + 405 629	38. 133 + 898 1 031	39. 761 + 638 1 399	40. 851 + 203 1 054	41. 161 + 876 1 037	42. 431 + 977 1 408

Find the sum

Complete all the activities (Addition).

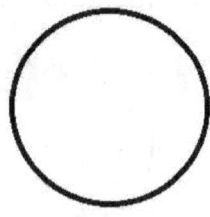

SCORE

1.	141 + 699	2.	559 + 338	3.	865 + 303	4.	874 + 597	5.	640 + 154	6.	418 + 744	7.	444 + 284

8.	442 + 525	9.	419 + 566	10.	996 + 448	11.	637 + 892	12.	230 + 974	13.	788 + 548	14.	937 + 589

15.	142 + 380	16.	768 + 601	17.	479 + 438	18.	896 + 611	19.	472 + 669	20.	280 + 975	21.	982 + 332

22.	143 + 170	23.	985 + 388	24.	554 + 577	25.	296 + 426	26.	784 + 594	27.	879 + 982	28.	440 + 248

29.	569 + 584	30.	549 + 417	31.	365 + 385	32.	555 + 360	33.	385 + 817	34.	365 + 106	35.	387 + 830

36.	153 + 749	37.	386 + 121	38.	466 + 405	39.	849 + 160	40.	987 + 500	41.	993 + 460	42.	393 + 715

Name:.................................... Date:....................................

Find the sum

Complete all the activities (Addition).

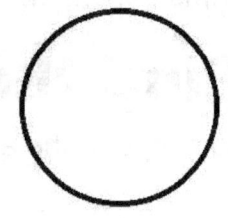

SCORE

1. 141 + 699 840	2. 559 + 338 897	3. 865 + 303 1 168	4. 874 + 597 1 471	5. 640 + 154 794	6. 418 + 744 1 162	7. 444 + 284 728
8. 442 + 525 967	9. 419 + 566 985	10. 996 + 448 1 444	11. 637 + 892 1 529	12. 230 + 974 1 204	13. 788 + 548 1 336	14. 937 + 589 1 526
15. 142 + 380 522	16. 768 + 601 1 369	17. 479 + 438 917	18. 896 + 611 1 507	19. 472 + 669 1 141	20. 280 + 975 1 255	21. 982 + 332 1 314
22. 143 + 170 313	23. 985 + 388 1 373	24. 554 + 577 1 131	25. 296 + 426 722	26. 784 + 594 1 378	27. 879 + 982 1 861	28. 440 + 248 688
29. 569 + 584 1 153	30. 549 + 417 966	31. 365 + 385 750	32. 555 + 360 915	33. 385 + 817 1 202	34. 365 + 106 471	35. 387 + 830 1 217
36. 153 + 749 902	37. 386 + 121 507	38. 466 + 405 871	39. 849 + 160 1 009	40. 987 + 500 1 487	41. 993 + 460 1 453	42. 393 + 715 1 108

Name:................................. Date:...............................

Find the sum

Complete all the activities (Addition).

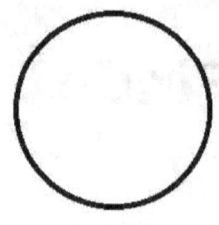

SCORE

1. 596 + 274	2. 721 + 496	3. 997 + 748	4. 142 + 887	5. 220 + 862	6. 710 + 696	7. 776 + 781
8. 784 + 269	9. 123 + 547	10. 290 + 465	11. 373 + 135	12. 308 + 857	13. 616 + 135	14. 398 + 509
15. 596 + 149	16. 423 + 904	17. 992 + 839	18. 764 + 211	19. 196 + 868	20. 108 + 637	21. 788 + 265
22. 629 + 524	23. 411 + 768	24. 628 + 168	25. 375 + 819	26. 478 + 546	27. 270 + 927	28. 442 + 561
29. 965 + 863	30. 879 + 963	31. 450 + 632	32. 391 + 171	33. 539 + 804	34. 704 + 936	35. 118 + 673
36. 173 + 253	37. 539 + 652	38. 676 + 208	39. 505 + 783	40. 547 + 847	41. 506 + 330	42. 234 + 776

Find the sum

Complete all the activities (Addition).

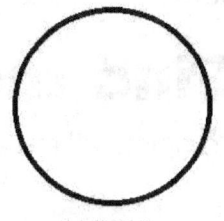

1. 596 + 274 870	2. 721 + 496 1 217	3. 997 + 748 1 745	4. 142 + 887 1 029	5. 220 + 862 1 082	6. 710 + 696 1 406	7. 776 + 781 1 557
8. 784 + 269 1 053	9. 123 + 547 670	10. 290 + 465 755	11. 373 + 135 508	12. 308 + 857 1 165	13. 616 + 135 751	14. 398 + 509 907
15. 596 + 149 745	16. 423 + 904 1 327	17. 992 + 839 1 831	18. 764 + 211 975	19. 196 + 868 1 064	20. 108 + 637 745	21. 788 + 265 1 053
22. 629 + 524 1 153	23. 411 + 768 1 179	24. 628 + 168 796	25. 375 + 819 1 194	26. 478 + 546 1 024	27. 270 + 927 1 197	28. 442 + 561 1 003
29. 965 + 863 1 828	30. 879 + 963 1 842	31. 450 + 632 1 082	32. 391 + 171 562	33. 539 + 804 1 343	34. 704 + 936 1 640	35. 118 + 673 791
36. 173 + 253 426	37. 539 + 652 1 191	38. 676 + 208 884	39. 505 + 783 1 288	40. 547 + 847 1 394	41. 506 + 330 836	42. 234 + 776 1 010

Find the sum

Complete all the activities (Addition).

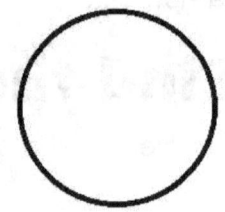

SCORE

1. 703 + 316	2. 580 + 663	3. 973 + 823	4. 920 + 197	5. 923 + 476	6. 687 + 837	7. 244 + 531
8. 718 + 308	9. 493 + 158	10. 343 + 142	11. 772 + 211	12. 651 + 748	13. 683 + 994	14. 396 + 227
15. 462 + 587	16. 243 + 736	17. 790 + 619	18. 734 + 163	19. 219 + 580	20. 327 + 509	21. 544 + 470
22. 836 + 310	23. 553 + 137	24. 673 + 800	25. 229 + 994	26. 421 + 217	27. 539 + 888	28. 469 + 592
29. 664 + 832	30. 979 + 978	31. 527 + 793	32. 733 + 436	33. 333 + 950	34. 147 + 624	35. 796 + 654
36. 189 + 912	37. 831 + 833	38. 596 + 402	39. 374 + 441	40. 294 + 646	41. 421 + 595	42. 997 + 467

Find the sum

Complete all the activities (Addition).

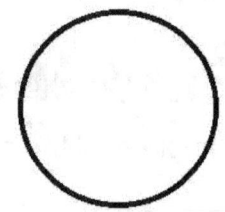

SCORE

1. 703 + 316 1 019	2. 580 + 663 1 243	3. 973 + 823 1 796	4. 920 + 197 1 117	5. 923 + 476 1 399	6. 687 + 837 1 524	7. 244 + 531 775
8. 718 + 308 1 026	9. 493 + 158 651	10. 343 + 142 485	11. 772 + 211 983	12. 651 + 748 1 399	13. 683 + 994 1 677	14. 396 + 227 623
15. 462 + 587 1 049	16. 243 + 736 979	17. 790 + 619 1 409	18. 734 + 163 897	19. 219 + 580 799	20. 327 + 509 836	21. 544 + 470 1 014
22. 836 + 310 1 146	23. 553 + 137 690	24. 673 + 800 1 473	25. 229 + 994 1 223	26. 421 + 217 638	27. 539 + 888 1 427	28. 469 + 592 1 061
29. 664 + 832 1 496	30. 979 + 978 1 957	31. 527 + 793 1 320	32. 733 + 436 1 169	33. 333 + 950 1 283	34. 147 + 624 771	35. 796 + 654 1 450
36. 189 + 912 1 101	37. 831 + 833 1 664	38. 596 + 402 998	39. 374 + 441 815	40. 294 + 646 940	41. 421 + 595 1 016	42. 997 + 467 1 464

Find the sum

Complete all the activities (Addition).

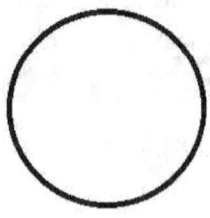

SCORE

1. 266 + 125	2. 297 + 901	3. 250 + 242	4. 499 + 386	5. 600 + 347	6. 981 + 750	7. 736 + 500
8. 976 + 602	9. 989 + 149	10. 104 + 574	11. 905 + 508	12. 386 + 704	13. 392 + 167	14. 318 + 323
15. 637 + 433	16. 657 + 931	17. 795 + 366	18. 270 + 308	19. 753 + 245	20. 474 + 654	21. 256 + 801
22. 413 + 583	23. 851 + 845	24. 665 + 873	25. 464 + 691	26. 656 + 372	27. 591 + 523	28. 561 + 757
29. 739 + 592	30. 609 + 183	31. 923 + 276	32. 748 + 267	33. 689 + 150	34. 953 + 469	35. 330 + 769
36. 848 + 844	37. 674 + 449	38. 418 + 204	39. 195 + 801	40. 852 + 546	41. 501 + 444	42. 858 + 945

Find the sum

Name:................................ Date:................................

Complete all the activities (Addition).

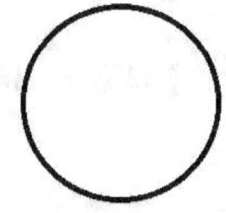

SCORE

1. 266 + 125 391	2. 297 + 901 1 198	3. 250 + 242 492	4. 499 + 386 885	5. 600 + 347 947	6. 981 + 750 1 731	7. 736 + 500 1 236
8. 976 + 602 1 578	9. 989 + 149 1 138	10. 104 + 574 678	11. 905 + 508 1 413	12. 386 + 704 1 090	13. 392 + 167 559	14. 318 + 323 641
15. 637 + 433 1 070	16. 657 + 931 1 588	17. 795 + 366 1 161	18. 270 + 308 578	19. 753 + 245 998	20. 474 + 654 1 128	21. 256 + 801 1 057
22. 413 + 583 996	23. 851 + 845 1 696	24. 665 + 873 1 538	25. 464 + 691 1 155	26. 656 + 372 1 028	27. 591 + 523 1 114	28. 561 + 757 1 318
29. 739 + 592 1 331	30. 609 + 183 792	31. 923 + 276 1 199	32. 748 + 267 1 015	33. 689 + 150 839	34. 953 + 469 1 422	35. 330 + 769 1 099
36. 848 + 844 1 692	37. 674 + 449 1 123	38. 418 + 204 622	39. 195 + 801 996	40. 852 + 546 1 398	41. 501 + 444 945	42. 858 + 945 1 803

Find the sum

Complete all the activities (Addition).

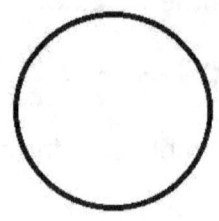

SCORE

| 1. | 776
+ 385 | 2. | 440
+ 372 | 3. | 804
+ 512 | 4. | 568
+ 262 | 5. | 636
+ 391 | 6. | 729
+ 979 | 7. | 153
+ 245 |

| 8. | 893
+ 802 | 9. | 891
+ 296 | 10. | 643
+ 694 | 11. | 843
+ 811 | 12. | 697
+ 242 | 13. | 205
+ 457 | 14. | 661
+ 937 |

| 15. | 739
+ 163 | 16. | 681
+ 580 | 17. | 417
+ 803 | 18. | 490
+ 404 | 19. | 709
+ 681 | 20. | 664
+ 817 | 21. | 301
+ 686 |

| 22. | 308
+ 382 | 23. | 718
+ 438 | 24. | 150
+ 949 | 25. | 907
+ 150 | 26. | 654
+ 306 | 27. | 800
+ 264 | 28. | 202
+ 964 |

| 29. | 687
+ 297 | 30. | 267
+ 447 | 31. | 850
+ 630 | 32. | 522
+ 573 | 33. | 710
+ 250 | 34. | 662
+ 349 | 35. | 754
+ 589 |

| 36. | 193
+ 310 | 37. | 630
+ 825 | 38. | 493
+ 257 | 39. | 603
+ 415 | 40. | 523
+ 766 | 41. | 407
+ 712 | 42. | 129
+ 689 |

Find the sum

Complete all the activities (Addition).

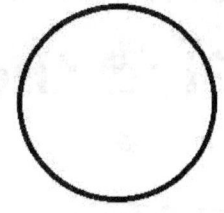

SCORE

1. 776 + 385 1 161	2. 440 + 372 812	3. 804 + 512 1 316	4. 568 + 262 830	5. 636 + 391 1 027	6. 729 + 979 1 708	7. 153 + 245 398
8. 893 + 802 1 695	9. 891 + 296 1 187	10. 643 + 694 1 337	11. 843 + 811 1 654	12. 697 + 242 939	13. 205 + 457 662	14. 661 + 937 1 598
15. 739 + 163 902	16. 681 + 580 1 261	17. 417 + 803 1 220	18. 490 + 404 894	19. 709 + 681 1 390	20. 664 + 817 1 481	21. 301 + 686 987
22. 308 + 382 690	23. 718 + 438 1 156	24. 150 + 949 1 099	25. 907 + 150 1 057	26. 654 + 306 960	27. 800 + 264 1 064	28. 202 + 964 1 166
29. 687 + 297 984	30. 267 + 447 714	31. 850 + 630 1 480	32. 522 + 573 1 095	33. 710 + 250 960	34. 662 + 349 1 011	35. 754 + 589 1 343
36. 193 + 310 503	37. 630 + 825 1 455	38. 493 + 257 750	39. 603 + 415 1 018	40. 523 + 766 1 289	41. 407 + 712 1 119	42. 129 + 689 818

Name:.............................. Date:..............................

Find the sum

Complete all the activities (Addition).

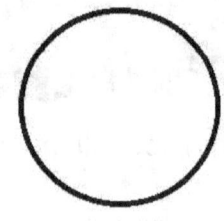

SCORE

1. 508 + 475	2. 778 + 253	3. 701 + 526	4. 716 + 511	5. 515 + 318	6. 966 + 669	7. 265 + 735
8. 671 + 569	9. 775 + 800	10. 295 + 418	11. 956 + 444	12. 674 + 974	13. 770 + 944	14. 461 + 372
15. 773 + 499	16. 140 + 238	17. 601 + 832	18. 165 + 335	19. 470 + 510	20. 648 + 328	21. 873 + 680
22. 794 + 413	23. 858 + 823	24. 991 + 867	25. 743 + 953	26. 991 + 606	27. 517 + 711	28. 732 + 164
29. 800 + 803	30. 482 + 756	31. 165 + 444	32. 737 + 115	33. 278 + 827	34. 168 + 432	35. 670 + 437
36. 198 + 365	37. 252 + 966	38. 544 + 376	39. 649 + 362	40. 215 + 946	41. 952 + 593	42. 628 + 819

Find the sum

Complete all the activities (Addition).

Name:.............................. Date:...............................

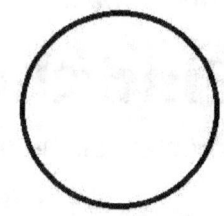

SCORE

1. 508	2. 778	3. 701	4. 716	5. 515	6. 966	7. 265
+ 475	+ 253	+ 526	+ 511	+ 318	+ 669	+ 735
983	1 031	1 227	1 227	833	1 635	1 000

8. 671	9. 775	10. 295	11. 956	12. 674	13. 770	14. 461
+ 569	+ 800	+ 418	+ 444	+ 974	+ 944	+ 372
1 240	1 575	713	1 400	1 648	1 714	833

15. 773	16. 140	17. 601	18. 165	19. 470	20. 648	21. 873
+ 499	+ 238	+ 832	+ 335	+ 510	+ 328	+ 680
1 272	378	1 433	500	980	976	1 553

22. 794	23. 858	24. 991	25. 743	26. 991	27. 517	28. 732
+ 413	+ 823	+ 867	+ 953	+ 606	+ 711	+ 164
1 207	1 681	1 858	1 696	1 597	1 228	896

29. 800	30. 482	31. 165	32. 737	33. 278	34. 168	35. 670
+ 803	+ 756	+ 444	+ 115	+ 827	+ 432	+ 437
1 603	1 238	609	852	1 105	600	1 107

36. 198	37. 252	38. 544	39. 649	40. 215	41. 952	42. 628
+ 365	+ 966	+ 376	+ 362	+ 946	+ 593	+ 819
563	1 218	920	1 011	1 161	1 545	1 447

Find the sum

Name:.................................   Date:.................................

Complete all the activities (Addition).

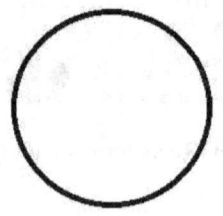

SCORE

| 1. | 508
+ 533 | 2. | 721
+ 889 | 3. | 200
+ 101 | 4. | 534
+ 666 | 5. | 128
+ 891 | 6. | 264
+ 529 | 7. | 942
+ 779 |

| 8. | 235
+ 131 | 9. | 754
+ 659 | 10. | 297
+ 406 | 11. | 855
+ 725 | 12. | 454
+ 550 | 13. | 267
+ 456 | 14. | 514
+ 538 |

| 15. | 688
+ 535 | 16. | 802
+ 570 | 17. | 815
+ 917 | 18. | 607
+ 744 | 19. | 207
+ 278 | 20. | 331
+ 670 | 21. | 902
+ 531 |

| 22. | 337
+ 360 | 23. | 802
+ 115 | 24. | 432
+ 869 | 25. | 227
+ 184 | 26. | 292
+ 393 | 27. | 397
+ 321 | 28. | 963
+ 781 |

| 29. | 179
+ 271 | 30. | 723
+ 743 | 31. | 586
+ 995 | 32. | 977
+ 592 | 33. | 189
+ 692 | 34. | 712
+ 125 | 35. | 272
+ 490 |

| 36. | 924
+ 576 | 37. | 428
+ 626 | 38. | 504
+ 424 | 39. | 488
+ 193 | 40. | 747
+ 251 | 41. | 124
+ 644 | 42. | 412
+ 604 |

Find the sum

Complete all the activities (Addition).

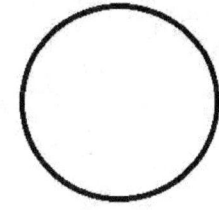

SCORE

| 1. | 508
+ 533
1 041 | 2. | 721
+ 889
1 610 | 3. | 200
+ 101
301 | 4. | 534
+ 666
1 200 | 5. | 128
+ 891
1 019 | 6. | 264
+ 529
793 | 7. | 942
+ 779
1 721 |

| 8. | 235
+ 131
366 | 9. | 754
+ 659
1 413 | 10. | 297
+ 406
703 | 11. | 855
+ 725
1 580 | 12. | 454
+ 550
1 004 | 13. | 267
+ 456
723 | 14. | 514
+ 538
1 052 |

| 15. | 688
+ 535
1 223 | 16. | 802
+ 570
1 372 | 17. | 815
+ 917
1 732 | 18. | 607
+ 744
1 351 | 19. | 207
+ 278
485 | 20. | 331
+ 670
1 001 | 21. | 902
+ 531
1 433 |

| 22. | 337
+ 360
697 | 23. | 802
+ 115
917 | 24. | 432
+ 869
1 301 | 25. | 227
+ 184
411 | 26. | 292
+ 393
685 | 27. | 397
+ 321
718 | 28. | 963
+ 781
1 744 |

| 29. | 179
+ 271
450 | 30. | 723
+ 743
1 466 | 31. | 586
+ 995
1 581 | 32. | 977
+ 592
1 569 | 33. | 189
+ 692
881 | 34. | 712
+ 125
837 | 35. | 272
+ 490
762 |

| 36. | 924
+ 576
1 500 | 37. | 428
+ 626
1 054 | 38. | 504
+ 424
928 | 39. | 488
+ 193
681 | 40. | 747
+ 251
998 | 41. | 124
+ 644
768 | 42. | 412
+ 604
1 016 |

Section 2

Name:................................   Date:...................................

Find the difference

Complete all the activities (Subtraction)

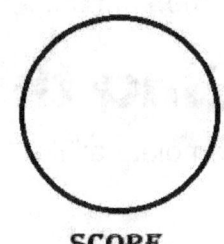

SCORE

1. 622 - 388	2. 282 - 185	3. 359 - 294	4. 681 - 419	5. 497 - 263	6. 123 - 109	7. 867 - 395	8. 815 - 491
9. 194 - 174	10. 407 - 335	11. 406 - 270	12. 308 - 160	13. 377 - 213	14. 921 - 367	15. 105 - 101	16. 505 - 163
17. 318 - 137	18. 329 - 102	19. 155 - 127	20. 450 - 421	21. 456 - 373	22. 199 - 186	23. 677 - 385	24. 769 - 451
25. 105 - 103	26. 482 - 453	27. 883 - 800	28. 526 - 111	29. 389 - 364	30. 241 - 224	31. 851 - 341	32. 588 - 240
33. 964 - 852	34. 231 - 190	35. 217 - 192	36. 536 - 367	37. 455 - 250	38. 433 - 291	39. 501 - 245	40. 680 - 573
41. 697 - 332	42. 387 - 188	43. 756 - 446	44. 356 - 189	45. 725 - 243	46. 719 - 191	47. 402 - 254	48. 174 - 106

Name:.................................   Date:.................................

Find the difference

Complete all the activities (Subtraction)

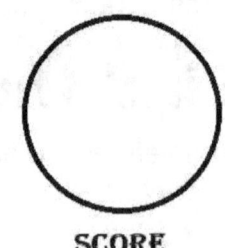

SCORE

1.	622	2.	282	3.	359	4.	681	5.	497	6.	123	7.	867	8.	815
	- 388		- 185		- 294		- 419		- 263		- 109		- 395		- 491
	234		97		65		262		234		14		472		324

9.	194	10.	407	11.	406	12.	308	13.	377	14.	921	15.	105	16.	505
	- 174		- 335		- 270		- 160		- 213		- 367		- 101		- 163
	20		72		136		148		164		554		4		342

17.	318	18.	329	19.	155	20.	450	21.	456	22.	199	23.	677	24.	769
	- 137		- 102		- 127		- 421		- 373		- 186		- 385		- 451
	181		227		28		29		83		13		292		318

25.	105	26.	482	27.	883	28.	526	29.	389	30.	241	31.	851	32.	588
	- 103		- 453		- 800		- 111		- 364		- 224		- 341		- 240
	2		29		83		415		25		17		510		348

33.	964	34.	231	35.	217	36.	536	37.	455	38.	433	39.	501	40.	680
	- 852		- 190		- 192		- 367		- 250		- 291		- 245		- 573
	112		41		25		169		205		142		256		107

41.	697	42.	387	43.	756	44.	356	45.	725	46.	719	47.	402	48.	174
	- 332		- 188		- 446		- 189		- 243		- 191		- 254		- 106
	365		199		310		167		482		528		148		68

Name:................................   Date:.................................

Find the difference

Complete all the activities (Subtraction)

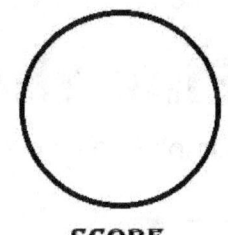

SCORE

1. 902 - 378	2. 355 - 278	3. 728 - 209	4. 530 - 497	5. 755 - 316	6. 568 - 198	7. 618 - 255	8. 632 - 115
9. 499 - 473	10. 114 - 112	11. 129 - 126	12. 410 - 115	13. 967 - 388	14. 963 - 824	15. 773 - 373	16. 201 - 144
17. 385 - 173	18. 418 - 410	19. 169 - 131	20. 479 - 315	21. 943 - 655	22. 763 - 668	23. 655 - 267	24. 640 - 521
25. 857 - 487	26. 970 - 149	27. 895 - 839	28. 972 - 965	29. 102 - 100	30. 759 - 341	31. 292 - 223	32. 257 - 151
33. 295 - 117	34. 903 - 356	35. 950 - 626	36. 956 - 953	37. 561 - 307	38. 916 - 405	39. 246 - 171	40. 488 - 357
41. 757 - 218	42. 996 - 313	43. 495 - 122	44. 593 - 291	45. 357 - 193	46. 974 - 836	47. 742 - 136	48. 417 - 194

Name:...................................   Date:................................

Find the difference

Complete all the activities (Subtraction)

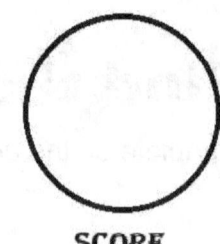

SCORE

| 1. | 902
- 378
524 | 2. | 355
- 278
77 | 3. | 728
- 209
519 | 4. | 530
- 497
33 | 5. | 755
- 316
439 | 6. | 568
- 198
370 | 7. | 618
- 255
363 | 8. | 632
- 115
517 |

| 9. | 499
- 473
26 | 10. | 114
- 112
2 | 11. | 129
- 126
3 | 12. | 410
- 115
295 | 13. | 967
- 388
579 | 14. | 963
- 824
139 | 15. | 773
- 373
400 | 16. | 201
- 144
57 |

| 17. | 385
- 173
212 | 18. | 418
- 410
8 | 19. | 169
- 131
38 | 20. | 479
- 315
164 | 21. | 943
- 655
288 | 22. | 763
- 668
95 | 23. | 655
- 267
388 | 24. | 640
- 521
119 |

| 25. | 857
- 487
370 | 26. | 970
- 149
821 | 27. | 895
- 839
56 | 28. | 972
- 965
7 | 29. | 102
- 100
2 | 30. | 759
- 341
418 | 31. | 292
- 223
69 | 32. | 257
- 151
106 |

| 33. | 295
- 117
178 | 34. | 903
- 356
547 | 35. | 950
- 626
324 | 36. | 956
- 953
3 | 37. | 561
- 307
254 | 38. | 916
- 405
511 | 39. | 246
- 171
75 | 40. | 488
- 357
131 |

| 41. | 757
- 218
539 | 42. | 996
- 313
683 | 43. | 495
- 122
373 | 44. | 593
- 291
302 | 45. | 357
- 193
164 | 46. | 974
- 836
138 | 47. | 742
- 136
606 | 48. | 417
- 194
223 |

Find the difference

Complete all the activities (Subtraction)

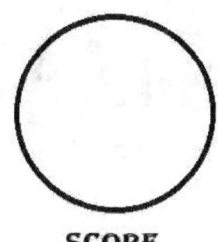

1. 535 − 281	2. 675 − 404	3. 467 − 203	4. 849 − 268	5. 603 − 588	6. 589 − 313	7. 120 − 112	8. 846 − 483
9. 822 − 151	10. 671 − 119	11. 679 − 642	12. 894 − 227	13. 586 − 346	14. 130 − 118	15. 681 − 164	16. 354 − 216
17. 187 − 153	18. 901 − 698	19. 863 − 790	20. 329 − 203	21. 412 − 341	22. 710 − 386	23. 560 − 152	24. 428 − 326
25. 578 − 319	26. 136 − 123	27. 478 − 221	28. 688 − 421	29. 579 − 330	30. 988 − 901	31. 323 − 156	32. 619 − 334
33. 719 − 154	34. 404 − 225	35. 734 − 153	36. 697 − 637	37. 833 − 236	38. 717 − 638	39. 333 − 197	40. 683 − 349
41. 227 − 144	42. 148 − 109	43. 163 − 140	44. 133 − 126	45. 959 − 396	46. 293 − 171	47. 924 − 390	48. 281 − 118

Name:.................................   Date:.................................

Find the difference

Complete all the activities (Subtraction)

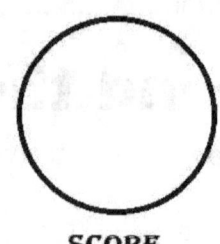

SCORE

1.	535 - 281 254	2.	675 - 404 271	3.	467 - 203 264	4.	849 - 268 581	5.	603 - 588 15	6.	589 - 313 276	7.	120 - 112 8	8.	846 - 483 363
9.	822 - 151 671	10.	671 - 119 552	11.	679 - 642 37	12.	894 - 227 667	13.	586 - 346 240	14.	130 - 118 12	15.	681 - 164 517	16.	354 - 216 138
17.	187 - 153 34	18.	901 - 698 203	19.	863 - 790 73	20.	329 - 203 126	21.	412 - 341 71	22.	710 - 386 324	23.	560 - 152 408	24.	428 - 326 102
25.	578 - 319 259	26.	136 - 123 13	27.	478 - 221 257	28.	688 - 421 267	29.	579 - 330 249	30.	988 - 901 87	31.	323 - 156 167	32.	619 - 334 285
33.	719 - 154 565	34.	404 - 225 179	35.	734 - 153 581	36.	697 - 637 60	37.	833 - 236 597	38.	717 - 638 79	39.	333 - 197 136	40.	683 - 349 334
41.	227 - 144 83	42.	148 - 109 39	43.	163 - 140 23	44.	133 - 126 7	45.	959 - 396 563	46.	293 - 171 122	47.	924 - 390 534	48.	281 - 118 163

Find the difference

Complete all the activities (Subtraction)

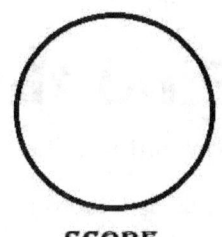
Name:.................................   Date:.................................

1. 275 - 140 --------	2. 801 - 398 --------	3. 796 - 763 --------	4. 985 - 496 --------	5. 814 - 201 --------	6. 845 - 275 --------	7. 669 - 545 --------	8. 359 - 130 --------
9. 847 - 519 --------	10. 784 - 719 --------	11. 277 - 169 --------	12. 442 - 117 --------	13. 145 - 125 --------	14. 510 - 248 --------	15. 881 - 330 --------	16. 983 - 509 --------
17. 985 - 621 --------	18. 708 - 119 --------	19. 671 - 228 --------	20. 687 - 211 --------	21. 422 - 295 --------	22. 454 - 429 --------	23. 827 - 683 --------	24. 229 - 122 --------
25. 151 - 105 --------	26. 584 - 272 --------	27. 631 - 257 --------	28. 910 - 753 --------	29. 862 - 655 --------	30. 437 - 172 --------	31. 545 - 377 --------	32. 620 - 360 --------
33. 242 - 152 --------	34. 326 - 110 --------	35. 196 - 125 --------	36. 533 - 363 --------	37. 916 - 566 --------	38. 592 - 391 --------	39. 995 - 504 --------	40. 317 - 154 --------
41. 476 - 207 --------	42. 374 - 312 --------	43. 780 - 526 --------	44. 884 - 639 --------	45. 718 - 638 --------	46. 265 - 250 --------	47. 863 - 436 --------	48. 724 - 559 --------

Name:................................   Date:................................

Find the difference

Complete all the activities (Subtraction)

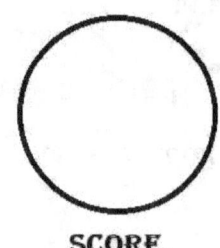

SCORE

| 1. | 275
 - 140
 135 | 2. | 801
 - 398
 403 | 3. | 796
 - 763
 33 | 4. | 985
 - 496
 489 | 5. | 814
 - 201
 613 | 6. | 845
 - 275
 570 | 7. | 669
 - 545
 124 | 8. | 359
 - 130
 229 |

9.   847   10.   784   11.   277   12.   442   13.   145   14.   510   15.   881   16.   983
    - 519   - 719   - 169   - 117   - 125   - 248   - 330   - 509
      328   65   108   325   20   262   551   474

17.   985   18.   708   19.   671   20.   687   21.   422   22.   454   23.   827   24.   229
    - 621   - 119   - 228   - 211   - 295   - 429   - 683   - 122
      364   589   443   476   127   25   144   107

25.   151   26.   584   27.   631   28.   910   29.   862   30.   437   31.   545   32.   620
    - 105   - 272   - 257   - 753   - 655   - 172   - 377   - 360
       46   312   374   157   207   265   168   260

33.   242   34.   326   35.   196   36.   533   37.   916   38.   592   39.   995   40.   317
    - 152   - 110   - 125   - 363   - 566   - 391   - 504   - 154
       90   216   71   170   350   201   491   163

41.   476   42.   374   43.   780   44.   884   45.   718   46.   265   47.   863   48.   724
    - 207   - 312   - 526   - 639   - 638   - 250   - 436   - 559
      269   62   254   245   80   15   427   165

© KingSchool Edition

Find the difference

Complete all the activities (Subtraction)

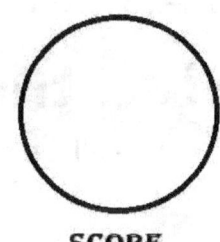

SCORE

1. 124 - 121 -------	2. 793 - 407 -------	3. 595 - 149 -------	4. 160 - 148 -------	5. 653 - 615 -------	6. 503 - 108 -------	7. 774 - 309 -------	8. 829 - 389 -------
9. 538 - 155 -------	10. 814 - 251 -------	11. 386 - 331 -------	12. 768 - 455 -------	13. 543 - 349 -------	14. 420 - 176 -------	15. 166 - 138 -------	16. 781 - 419 -------
17. 888 - 740 -------	18. 608 - 151 -------	19. 711 - 587 -------	20. 728 - 300 -------	21. 295 - 199 -------	22. 241 - 216 -------	23. 260 - 236 -------	24. 944 - 419 -------
25. 154 - 142 -------	26. 477 - 373 -------	27. 680 - 380 -------	28. 475 - 183 -------	29. 237 - 133 -------	30. 954 - 943 -------	31. 768 - 654 -------	32. 752 - 126 -------
33. 256 - 253 -------	34. 228 - 220 -------	35. 770 - 576 -------	36. 745 - 545 -------	37. 484 - 283 -------	38. 994 - 826 -------	39. 683 - 635 -------	40. 417 - 303 -------
41. 662 - 650 -------	42. 259 - 146 -------	43. 241 - 105 -------	44. 839 - 392 -------	45. 455 - 222 -------	46. 953 - 476 -------	47. 862 - 641 -------	48. 617 - 300 -------

Name:................................ Date:..............................

Find the difference

Complete all the activities (Subtraction)

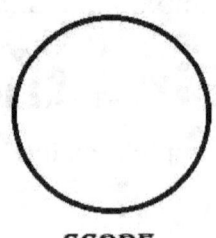

SCORE

1.	124	2.	793	3.	595	4.	160	5.	653	6.	503	7.	774	8.	829
	- 121		- 407		- 149		- 148		- 615		- 108		- 309		- 389
	3		386		446		12		38		395		465		440

9.	538	10.	814	11.	386	12.	768	13.	543	14.	420	15.	166	16.	781
	- 155		- 251		- 331		- 455		- 349		- 176		- 138		- 419
	383		563		55		313		194		244		28		362

17.	888	18.	608	19.	711	20.	728	21.	295	22.	241	23.	260	24.	944
	- 740		- 151		- 587		- 300		- 199		- 216		- 236		- 419
	148		457		124		428		96		25		24		525

25.	154	26.	477	27.	680	28.	475	29.	237	30.	954	31.	768	32.	752
	- 142		- 373		- 380		- 183		- 133		- 943		- 654		- 126
	12		104		300		292		104		11		114		626

33.	256	34.	228	35.	770	36.	745	37.	484	38.	994	39.	683	40.	417
	- 253		- 220		- 576		- 545		- 283		- 826		- 635		- 303
	3		8		194		200		201		168		48		114

41.	662	42.	259	43.	241	44.	839	45.	455	46.	953	47.	862	48.	617
	- 650		- 146		- 105		- 392		- 222		- 476		- 641		- 300
	12		113		136		447		233		477		221		317

Name:................................   Date:................................

Find the difference

Complete all the activities (Subtraction)

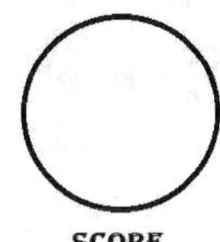

SCORE

1. 801 - 183 --------	2. 685 - 216 --------	3. 738 - 100 --------	4. 669 - 299 --------	5. 162 - 145 --------	6. 366 - 145 --------	7. 538 - 365 --------	8. 613 - 220 --------
9. 897 - 347 --------	10. 426 - 349 --------	11. 687 - 357 --------	12. 893 - 211 --------	13. 242 - 206 --------	14. 588 - 295 --------	15. 785 - 290 --------	16. 858 - 644 --------
17. 910 - 592 --------	18. 129 - 115 --------	19. 340 - 137 --------	20. 937 - 623 --------	21. 410 - 356 --------	22. 297 - 184 --------	23. 764 - 534 --------	24. 915 - 744 --------
25. 795 - 596 --------	26. 598 - 368 --------	27. 413 - 354 --------	28. 760 - 326 --------	29. 312 - 208 --------	30. 136 - 135 --------	31. 511 - 196 --------	32. 629 - 519 --------
33. 897 - 561 --------	34. 341 - 312 --------	35. 614 - 121 --------	36. 970 - 737 --------	37. 441 - 352 --------	38. 478 - 412 --------	39. 297 - 190 --------	40. 461 - 282 --------
41. 856 - 669 --------	42. 664 - 127 --------	43. 750 - 298 --------	44. 432 - 130 --------	45. 885 - 642 --------	46. 365 - 300 --------	47. 203 - 200 --------	48. 250 - 117 --------

Name:................................   Date:.................................

Find the difference

Complete all the activities (Subtraction)

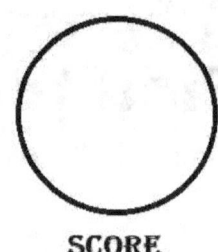

SCORE

1. 801 - 183 618	2. 685 - 216 469	3. 738 - 100 638	4. 669 - 299 370	5. 162 - 145 17	6. 366 - 145 221	7. 538 - 365 173	8. 613 - 220 393
9. 897 - 347 550	10. 426 - 349 77	11. 687 - 357 330	12. 893 - 211 682	13. 242 - 206 36	14. 588 - 295 293	15. 785 - 290 495	16. 858 - 644 214
17. 910 - 592 318	18. 129 - 115 14	19. 340 - 137 203	20. 937 - 623 314	21. 410 - 356 54	22. 297 - 184 113	23. 764 - 534 230	24. 915 - 744 171
25. 795 - 596 199	26. 598 - 368 230	27. 413 - 354 59	28. 760 - 326 434	29. 312 - 208 104	30. 136 - 135 1	31. 511 - 196 315	32. 629 - 519 110
33. 897 - 561 336	34. 341 - 312 29	35. 614 - 121 493	36. 970 - 737 233	37. 441 - 352 89	38. 478 - 412 66	39. 297 - 190 107	40. 461 - 282 179
41. 856 - 669 187	42. 664 - 127 537	43. 750 - 298 452	44. 432 - 130 302	45. 885 - 642 243	46. 365 - 300 65	47. 203 - 200 3	48. 250 - 117 133

Name:................................ Date:................................

Find the difference

Complete all the activities (Subtraction)

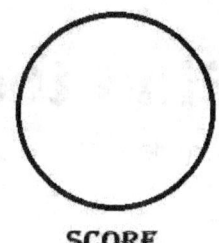

SCORE

1.	314	2.	587	3.	447	4.	296	5.	771	6.	243	7.	450	8.	211
	- 191		- 178		- 109		- 244		- 652		- 187		- 418		- 187
	--------		--------		--------		--------		--------		--------		--------		--------

9.	199	10.	208	11.	645	12.	162	13.	289	14.	106	15.	770	16.	654
	- 153		- 176		- 395		- 136		- 133		- 105		- 242		- 327
	--------		--------		--------		--------		--------		--------		--------		--------

17.	613	18.	629	19.	920	20.	207	21.	580	22.	371	23.	232	24.	731
	- 258		- 519		- 298		- 187		- 529		- 344		- 200		- 709
	--------		--------		--------		--------		--------		--------		--------		--------

25.	610	26.	680	27.	863	28.	555	29.	422	30.	954	31.	117	32.	988
	- 362		- 436		- 453		- 417		- 409		- 279		- 106		- 574
	--------		--------		--------		--------		--------		--------		--------		--------

33.	285	34.	135	35.	387	36.	881	37.	475	38.	253	39.	315	40.	490
	- 239		- 103		- 195		- 779		- 214		- 209		- 136		- 276
	--------		--------		--------		--------		--------		--------		--------		--------

41.	527	42.	382	43.	610	44.	604	45.	821	46.	490	47.	362	48.	815
	- 118		- 165		- 119		- 155		- 742		- 471		- 338		- 241
	--------		--------		--------		--------		--------		--------		--------		--------

Name:................................ Date:................................

Find the difference

Complete all the activities (Subtraction)

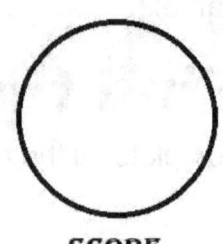

SCORE

| 1. | 314
- 191
123 | 2. | 587
- 178
409 | 3. | 447
- 109
338 | 4. | 296
- 244
52 | 5. | 771
- 652
119 | 6. | 243
- 187
56 | 7. | 450
- 418
32 | 8. | 211
- 187
24 |

| 9. | 199
- 153
46 | 10. | 208
- 176
32 | 11. | 645
- 395
250 | 12. | 162
- 136
26 | 13. | 289
- 133
156 | 14. | 106
- 105
1 | 15. | 770
- 242
528 | 16. | 654
- 327
327 |

| 17. | 613
- 258
355 | 18. | 629
- 519
110 | 19. | 920
- 298
622 | 20. | 207
- 187
20 | 21. | 580
- 529
51 | 22. | 371
- 344
27 | 23. | 232
- 200
32 | 24. | 731
- 709
22 |

| 25. | 610
- 362
248 | 26. | 680
- 436
244 | 27. | 863
- 453
410 | 28. | 555
- 417
138 | 29. | 422
- 409
13 | 30. | 954
- 279
675 | 31. | 117
- 106
11 | 32. | 988
- 574
414 |

| 33. | 285
- 239
46 | 34. | 135
- 103
32 | 35. | 387
- 195
192 | 36. | 881
- 779
102 | 37. | 475
- 214
261 | 38. | 253
- 209
44 | 39. | 315
- 136
179 | 40. | 490
- 276
214 |

| 41. | 527
- 118
409 | 42. | 382
- 165
217 | 43. | 610
- 119
491 | 44. | 604
- 155
449 | 45. | 821
- 742
79 | 46. | 490
- 471
19 | 47. | 362
- 338
24 | 48. | 815
- 241
574 |

Name:............................   Date:.............................

Find the difference

Complete all the activities (Subtraction)

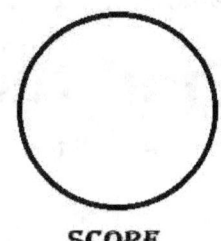

SCORE

1. 914 − 725	2. 726 − 349	3. 747 − 501	4. 681 − 237	5. 680 − 550
6. 246 − 179	7. 363 − 309	8. 512 − 179		

1. 914
  - 725
  ----

2. 726
  - 349
  ----

3. 747
  - 501
  ----

4. 681
  - 237
  ----

5. 680
  - 550
  ----

6. 246
  - 179
  ----

7. 363
  - 309
  ----

8. 512
  - 179
  ----

9. 574
  - 337
  ----

10. 760
  - 746
  ----

11. 121
  - 116
  ----

12. 968
  - 246
  ----

13. 270
  - 258
  ----

14. 351
  - 208
  ----

15. 907
  - 840
  ----

16. 619
  - 363
  ----

17. 558
  - 546
  ----

18. 824
  - 441
  ----

19. 524
  - 115
  ----

20. 757
  - 662
  ----

21. 456
  - 127
  ----

22. 627
  - 517
  ----

23. 886
  - 113
  ----

24. 833
  - 105
  ----

25. 488
  - 194
  ----

26. 482
  - 403
  ----

27. 258
  - 229
  ----

28. 256
  - 123
  ----

29. 821
  - 481
  ----

30. 288
  - 248
  ----

31. 406
  - 166
  ----

32. 405
  - 322
  ----

33. 278
  - 217
  ----

34. 131
  - 110
  ----

35. 989
  - 978
  ----

36. 747
  - 417
  ----

37. 791
  - 707
  ----

38. 624
  - 596
  ----

39. 253
  - 127
  ----

40. 364
  - 146
  ----

41. 203
  - 128
  ----

42. 997
  - 687
  ----

43. 543
  - 448
  ----

44. 398
  - 306
  ----

45. 819
  - 572
  ----

46. 701
  - 667
  ----

47. 844
  - 800
  ----

48. 563
  - 183

Find the difference

Complete all the activities (Subtraction)

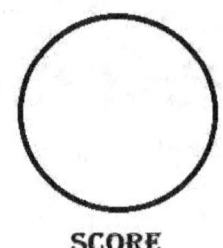

SCORE

Name:................................. Date:................................

1. 914 - 725 189	2. 726 - 349 377	3. 747 - 501 246	4. 681 - 237 444	5. 680 - 550 130	6. 246 - 179 67	7. 363 - 309 54	8. 512 - 179 333
9. 574 - 337 237	10. 760 - 746 14	11. 121 - 116 5	12. 968 - 246 722	13. 270 - 258 12	14. 351 - 208 143	15. 907 - 840 67	16. 619 - 363 256
17. 558 - 546 12	18. 824 - 441 383	19. 524 - 115 409	20. 757 - 662 95	21. 456 - 127 329	22. 627 - 517 110	23. 886 - 113 773	24. 833 - 105 728
25. 488 - 194 294	26. 482 - 403 79	27. 258 - 229 29	28. 256 - 123 133	29. 821 - 481 340	30. 288 - 248 40	31. 406 - 166 240	32. 405 - 322 83
33. 278 - 217 61	34. 131 - 110 21	35. 989 - 978 11	36. 747 - 417 330	37. 791 - 707 84	38. 624 - 596 28	39. 253 - 127 126	40. 364 - 146 218
41. 203 - 128 75	42. 997 - 687 310	43. 543 - 448 95	44. 398 - 306 92	45. 819 - 572 247	46. 701 - 667 34	47. 844 - 800 44	48. 563 - 183 380

Find the difference

Complete all the activities (Subtraction)

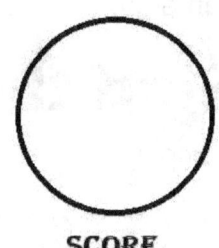

SCORE

1.	763	2.	633	3.	983	4.	116	5.	570	6.	121	7.	311	8.	980
	- 669		- 217		- 538		- 107		- 539		- 106		- 148		- 653

9.	311	10.	707	11.	297	12.	312	13.	391	14.	567	15.	387	16.	696
	- 108		- 141		- 235		- 188		- 200		- 255		- 323		- 593

17.	427	18.	288	19.	622	20.	272	21.	329	22.	218	23.	785	24.	511
	- 232		- 159		- 139		- 206		- 227		- 166		- 124		- 335

25.	710	26.	737	27.	339	28.	963	29.	660	30.	906	31.	978	32.	213
	- 125		- 604		- 207		- 145		- 201		- 615		- 702		- 184

33.	343	34.	938	35.	116	36.	143	37.	860	38.	831	39.	598	40.	686
	- 147		- 797		- 111		- 107		- 143		- 586		- 437		- 269

41.	896	42.	688	43.	161	44.	605	45.	499	46.	720	47.	260	48.	677
	- 854		- 263		- 153		- 563		- 162		- 348		- 188		- 428

Find the difference

Complete all the activities (Subtraction)

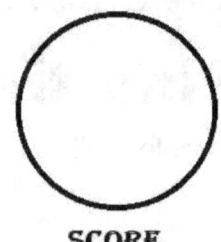

SCORE

1. 763 - 669 94	2. 633 - 217 416	3. 983 - 538 445	4. 116 - 107 9	5. 570 - 539 31	6. 121 - 106 15	7. 311 - 148 163	8. 980 - 653 327
9. 311 - 108 203	10. 707 - 141 566	11. 297 - 235 62	12. 312 - 188 124	13. 391 - 200 191	14. 567 - 255 312	15. 387 - 323 64	16. 696 - 593 103
17. 427 - 232 195	18. 288 - 159 129	19. 622 - 139 483	20. 272 - 206 66	21. 329 - 227 102	22. 218 - 166 52	23. 785 - 124 661	24. 511 - 335 176
25. 710 - 125 585	26. 737 - 604 133	27. 339 - 207 132	28. 963 - 145 818	29. 660 - 201 459	30. 906 - 615 291	31. 978 - 702 276	32. 213 - 184 29
33. 343 - 147 196	34. 938 - 797 141	35. 116 - 111 5	36. 143 - 107 36	37. 860 - 143 717	38. 831 - 586 245	39. 598 - 437 161	40. 686 - 269 417
41. 896 - 854 42	42. 688 - 263 425	43. 161 - 153 8	44. 605 - 563 42	45. 499 - 162 337	46. 720 - 348 372	47. 260 - 188 72	48. 677 - 428 249

Find the difference

Complete all the activities (Subtraction)

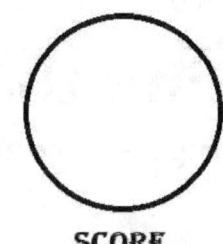

SCORE

1.   485
  - 248
	2. 358
	  - 246
		3. 192
		  - 158
			4. 598
			  - 106
				5. 993
				  - 589
					6. 950
					  - 278
						7. 670
						  - 268
							8. 548
							  - 523

9.   575
  - 389
	10. 778
	  - 120
		11. 129
		  - 125
			12. 711
			  - 343
				13. 669
				  - 109
					14. 449
					  - 401
						15. 977
						  - 668
							16. 984
							  - 360

17.   473
  - 192
	18. 313
	  - 136
		19. 451
		  - 130
			20. 267
			  - 143
				21. 453
				  - 327
					22. 356
					  - 233
						23. 934
						  - 394
							24. 355
							  - 138

25.   691
  - 541
	26. 884
	  - 546
		27. 816
		  - 539
			28. 582
			  - 380
				29. 734
				  - 664
					30. 939
					  - 173
						31. 314
						  - 248
							32. 132
							  - 111

33.   608
  - 431
	34. 815
	  - 378
		35. 153
		  - 117
			36. 599
			  - 322
				37. 480
				  - 288
					38. 672
					  - 323
						39. 605
						  - 181
							40. 523
							  - 219

41.   758
  - 258
	42. 284
	  - 142
		43. 797
		  - 388
			44. 547
			  - 458
				45. 738
				  - 518
					46. 517
					  - 309
						47. 236
						  - 113
							48. 119
							  - 113

Find the difference
Complete all the activities (Subtraction)

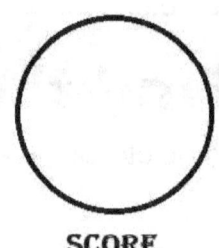

SCORE

1.	485	2.	358	3.	192	4.	598	5.	993	6.	950	7.	670	8.	548
	- 248		- 246		- 158		- 106		- 589		- 278		- 268		- 523
	237		112		34		492		404		672		402		25

9.	575	10.	778	11.	129	12.	711	13.	669	14.	449	15.	977	16.	984
	- 389		- 120		- 125		- 343		- 109		- 401		- 668		- 360
	186		658		4		368		560		48		309		624

17.	473	18.	313	19.	451	20.	267	21.	453	22.	356	23.	934	24.	355
	- 192		- 136		- 130		- 143		- 327		- 233		- 394		- 138
	281		177		321		124		126		123		540		217

25.	691	26.	884	27.	816	28.	582	29.	734	30.	939	31.	314	32.	132
	- 541		- 546		- 539		- 380		- 664		- 173		- 248		- 111
	150		338		277		202		70		766		66		21

33.	608	34.	815	35.	153	36.	599	37.	480	38.	672	39.	605	40.	523
	- 431		- 378		- 117		- 322		- 288		- 323		- 181		- 219
	177		437		36		277		192		349		424		304

41.	758	42.	284	43.	797	44.	547	45.	738	46.	517	47.	236	48.	119
	- 258		- 142		- 388		- 458		- 518		- 309		- 113		- 113
	500		142		409		89		220		208		123		6

Find the difference

Complete all the activities (Subtraction)

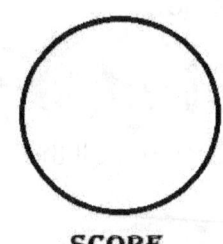

SCORE

1. 557 - 123	2. 713 - 478	3. 144 - 107	4. 362 - 117	5. 326 - 112	6. 237 - 197	7. 997 - 625	8. 588 - 573
9. 167 - 121	10. 105 - 103	11. 479 - 374	12. 267 - 146	13. 597 - 244	14. 304 - 256	15. 683 - 294	16. 121 - 103
17. 280 - 234	18. 279 - 249	19. 859 - 327	20. 330 - 277	21. 524 - 104	22. 264 - 221	23. 357 - 343	24. 326 - 310
25. 389 - 174	26. 103 - 101	27. 371 - 213	28. 908 - 735	29. 877 - 418	30. 157 - 143	31. 948 - 548	32. 475 - 444
33. 127 - 127	34. 558 - 217	35. 448 - 347	36. 212 - 187	37. 156 - 113	38. 196 - 140	39. 585 - 317	40. 988 - 639
41. 135 - 110	42. 497 - 256	43. 397 - 273	44. 868 - 702	45. 859 - 413	46. 970 - 479	47. 312 - 205	48. 844 - 802

Find the difference

Complete all the activities (Subtraction)

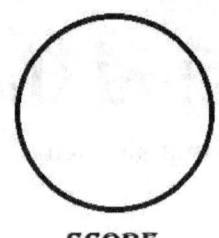

SCORE

1. 557 - 123 434	2. 713 - 478 235	3. 144 - 107 37	4. 362 - 117 245	5. 326 - 112 214	6. 237 - 197 40	7. 997 - 625 372	8. 588 - 573 15
9. 167 - 121 46	10. 105 - 103 2	11. 479 - 374 105	12. 267 - 146 121	13. 597 - 244 353	14. 304 - 256 48	15. 683 - 294 389	16. 121 - 103 18
17. 280 - 234 46	18. 279 - 249 30	19. 859 - 327 532	20. 330 - 277 53	21. 524 - 104 420	22. 264 - 221 43	23. 357 - 343 14	24. 326 - 310 16
25. 389 - 174 215	26. 103 - 101 2	27. 371 - 213 158	28. 908 - 735 173	29. 877 - 418 459	30. 157 - 143 14	31. 948 - 548 400	32. 475 - 444 31
33. 127 - 127 0	34. 558 - 217 341	35. 448 - 347 101	36. 212 - 187 25	37. 156 - 113 43	38. 196 - 140 56	39. 585 - 317 268	40. 988 - 639 349
41. 135 - 110 25	42. 497 - 256 241	43. 397 - 273 124	44. 868 - 702 166	45. 859 - 413 446	46. 970 - 479 491	47. 312 - 205 107	48. 844 - 802 42

Name:................................   Date:................................

Find the difference

Complete all the activities (Subtraction)

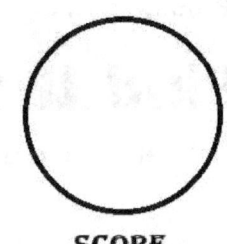

SCORE

1.	632 - 241 --------	2.	737 - 284 --------	3.	493 - 491 --------	4.	587 - 398 --------	5.	217 - 127 --------	6.	439 - 122 --------	7.	379 - 236 --------	8.	321 - 192 --------

9. 478 − 273

10. 748 − 518

11. 178 − 102

12. 629 − 336

13. 228 − 193

14. 469 − 460

15. 768 − 560

16. 586 − 375

17. 467 − 160

18. 838 − 303

19. 764 − 564

20. 252 − 106

21. 779 − 664

22. 337 − 124

23. 286 − 247

24. 526 − 253

25. 250 − 182

26. 114 − 111

27. 797 − 227

28. 870 − 538

29. 709 − 259

30. 232 − 194

31. 626 − 399

32. 880 − 444

33. 718 − 601

34. 692 − 187

35. 698 − 681

36. 711 − 541

37. 359 − 276

38. 900 − 305

39. 782 − 431

40. 316 − 176

41. 327 − 112

42. 826 − 241

43. 706 − 562

44. 811 − 261

45. 376 − 170

46. 637 − 356

47. 597 − 327

48. 511 − 177

Find the difference

Complete all the activities (Subtraction)

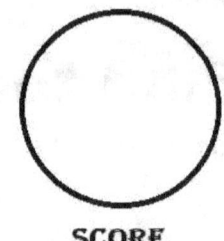

SCORE

| 1. | 632 − 241 = 391 | 2. | 737 − 284 = 453 | 3. | 493 − 491 = 2 | 4. | 587 − 398 = 189 | 5. | 217 − 127 = 90 | 6. | 439 − 122 = 317 | 7. | 379 − 236 = 143 | 8. | 321 − 192 = 129 |

| 9. | 478 − 273 = 205 | 10. | 748 − 518 = 230 | 11. | 178 − 102 = 76 | 12. | 629 − 336 = 293 | 13. | 228 − 193 = 35 | 14. | 469 − 460 = 9 | 15. | 768 − 560 = 208 | 16. | 586 − 375 = 211 |

| 17. | 467 − 160 = 307 | 18. | 838 − 303 = 535 | 19. | 764 − 564 = 200 | 20. | 252 − 106 = 146 | 21. | 779 − 664 = 115 | 22. | 337 − 124 = 213 | 23. | 286 − 247 = 39 | 24. | 526 − 253 = 273 |

| 25. | 250 − 182 = 68 | 26. | 114 − 111 = 3 | 27. | 797 − 227 = 570 | 28. | 870 − 538 = 332 | 29. | 709 − 259 = 450 | 30. | 232 − 194 = 38 | 31. | 626 − 399 = 227 | 32. | 880 − 444 = 436 |

| 33. | 718 − 601 = 117 | 34. | 692 − 187 = 505 | 35. | 698 − 681 = 17 | 36. | 711 − 541 = 170 | 37. | 359 − 276 = 83 | 38. | 900 − 305 = 595 | 39. | 782 − 431 = 351 | 40. | 316 − 176 = 140 |

| 41. | 327 − 112 = 215 | 42. | 826 − 241 = 585 | 43. | 706 − 562 = 144 | 44. | 811 − 261 = 550 | 45. | 376 − 170 = 206 | 46. | 637 − 356 = 281 | 47. | 597 − 327 = 270 | 48. | 511 − 177 = 334 |

Name:................................   Date:................................

Find the difference

Complete all the activities (Subtraction)

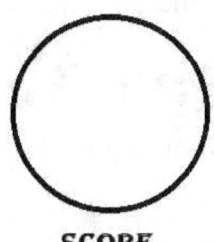

SCORE

1. 966	2. 473	3. 217	4. 723	5. 882	6. 540	7. 362	8. 828
- 813	- 248	- 138	- 221	- 330	- 142	- 257	- 239

9. 119	10. 514	11. 783	12. 789	13. 118	14. 854	15. 436	16. 251
- 115	- 320	- 560	- 737	- 108	- 411	- 168	- 226

17. 299	18. 364	19. 164	20. 302	21. 424	22. 392	23. 345	24. 940
- 240	- 347	- 154	- 287	- 299	- 211	- 270	- 694

25. 498	26. 509	27. 996	28. 485	29. 151	30. 438	31. 539	32. 837
- 262	- 411	- 929	- 211	- 125	- 336	- 223	- 592

33. 213	34. 887	35. 376	36. 208	37. 611	38. 255	39. 414	40. 686
- 174	- 564	- 341	- 119	- 373	- 123	- 221	- 291

41. 547	42. 972	43. 772	44. 851	45. 346	46. 890	47. 522	48. 493
- 159	- 944	- 374	- 323	- 101	- 559	- 322	- 131

Name:.................................... Date:....................................

Find the difference

Complete all the activities (Subtraction)

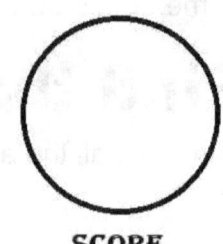

SCORE

1.   966
  - 813
  153
	2. 473
	  - 248
	  225
		3. 217
		  - 138
		  79
			4. 723
			  - 221
			  502
				5. 882
				  - 330
				  552
					6. 540
					  - 142
					  398
						7. 362
						  - 257
						  105
							8. 828
							  - 239
							  589

9.   119
  - 115
  4
	10. 514
	  - 320
	  194
		11. 783
		  - 560
		  223
			12. 789
			  - 737
			  52
				13. 118
				  - 108
				  10
					14. 854
					  - 411
					  443
						15. 436
						  - 168
						  268
							16. 251
							  - 226
							  25

17.   299
  - 240
  59
	18. 364
	  - 347
	  17
		19. 164
		  - 154
		  10
			20. 302
			  - 287
			  15
				21. 424
				  - 299
				  125
					22. 392
					  - 211
					  181
						23. 345
						  - 270
						  75
							24. 940
							  - 694
							  246

25.   498
  - 262
  236
	26. 509
	  - 411
	  98
		27. 996
		  - 929
		  67
			28. 485
			  - 211
			  274
				29. 151
				  - 125
				  26
					30. 438
					  - 336
					  102
						31. 539
						  - 223
						  316
							32. 837
							  - 592
							  245

33.   213
  - 174
  39
	34. 887
	  - 564
	  323
		35. 376
		  - 341
		  35
			36. 208
			  - 119
			  89
				37. 611
				  - 373
				  238
					38. 255
					  - 123
					  132
						39. 414
						  - 221
						  193
							40. 686
							  - 291
							  395

41.   547
  - 159
  388
	42. 972
	  - 944
	  28
		43. 772
		  - 374
		  398
			44. 851
			  - 323
			  528
				45. 346
				  - 101
				  245
					46. 890
					  - 559
					  331
						47. 522
						  - 322
						  200
							48. 493
							  - 131
							  362

Name:....................................   Date:....................................

Find the difference

Complete all the activities (Subtraction)

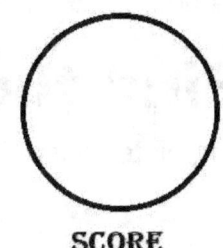

SCORE

1. 813 - 569	2. 328 - 240	3. 837 - 354	4. 460 - 305	5. 357 - 172	6. 273 - 180	7. 777 - 291	8. 945 - 666

--------   --------   --------   --------   --------   --------   --------   --------

9. 853   10. 279   11. 379   12. 927   13. 840   14. 385   15. 995   16. 736
  - 138   - 164   - 338   - 694   - 528   - 186   - 583   - 126

--------   --------   --------   --------   --------   --------   --------   --------

17. 188   18. 418   19. 219   20. 555   21. 758   22. 713   23. 294   24. 589
  - 108   - 320   - 104   - 170   - 364   - 387   - 172   - 585

--------   --------   --------   --------   --------   --------   --------   --------

25. 157   26. 768   27. 394   28. 791   29. 763   30. 812   31. 125   32. 615
  - 112   - 380   - 124   - 750   - 455   - 271   - 102   - 484

--------   --------   --------   --------   --------   --------   --------   --------

33. 614   34. 427   35. 935   36. 583   37. 589   38. 660   39. 253   40. 552
  - 178   - 316   - 307   - 487   - 123   - 609   - 169   - 303

--------   --------   --------   --------   --------   --------   --------   --------

41. 131   42. 315   43. 816   44. 741   45. 391   46. 958   47. 320   48. 213
  - 110   - 269   - 306   - 110   - 220   - 612   - 158   - 172

--------   --------   --------   --------   --------   --------   --------   --------

Name:................................   Date:................................

Find the difference

Complete all the activities (Subtraction)

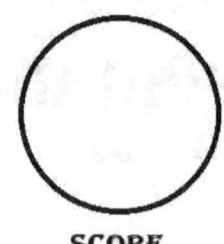

SCORE

1. 813
  - 569
  244
	2. 328
	  - 240
	  88
		3. 837
		  - 354
		  483
			4. 460
			  - 305
			  155
				5. 357
				  - 172
				  185
					6. 273
					  - 180
					  93
						7. 777
						  - 291
						  486
							8. 945
							  - 666
							  279

9. 853
  - 138
  715
	10. 279
	  - 164
	  115
		11. 379
		  - 338
		  41
			12. 927
			  - 694
			  233
				13. 840
				  - 528
				  312
					14. 385
					  - 186
					  199
						15. 995
						  - 583
						  412
							16. 736
							  - 126
							  610

17. 188
  - 108
  80
	18. 418
	  - 320
	  98
		19. 219
		  - 104
		  115
			20. 555
			  - 170
			  385
				21. 758
				  - 364
				  394
					22. 713
					  - 387
					  326
						23. 294
						  - 172
						  122
							24. 589
							  - 585
							  4

25. 157
  - 112
  45
	26. 768
	  - 380
	  388
		27. 394
		  - 124
		  270
			28. 791
			  - 750
			  41
				29. 763
				  - 455
				  308
					30. 812
					  - 271
					  541
						31. 125
						  - 102
						  23
							32. 615
							  - 484
							  131

33. 614
  - 178
  436
	34. 427
	  - 316
	  111
		35. 935
		  - 307
		  628
			36. 583
			  - 487
			  96
				37. 589
				  - 123
				  466
					38. 660
					  - 609
					  51
						39. 253
						  - 169
						  84
							40. 552
							  - 303
							  249

41. 131
  - 110
  21
	42. 315
	  - 269
	  46
		43. 816
		  - 306
		  510
			44. 741
			  - 110
			  631
				45. 391
				  - 220
				  171
					46. 958
					  - 612
					  346
						47. 320
						  - 158
						  162
							48. 213
							  - 172
							  41

Find the difference

Complete all the activities (Subtraction)

Name:................................ Date:...............................

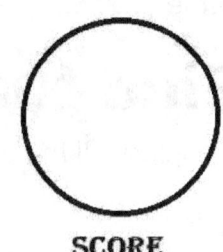

SCORE

1.   844
  - 796
	2. 907
	  - 151
		3. 109
		  - 108
			4. 156
			  - 146
				5. 704
				  - 669
					6. 887
					  - 321
						7. 512
						  - 454
							8. 793
							  - 299

9.   130
  - 110
	10. 285
	  - 230
		11. 871
		  - 706
			12. 954
			  - 342
				13. 494
				  - 237
					14. 402
					  - 256
						15. 364
						  - 348
							16. 600
							  - 239

17.   928
  - 918
	18. 174
	  - 117
		19. 598
		  - 319
			20. 495
			  - 160
				21. 336
				  - 200
					22. 204
					  - 189
						23. 471
						  - 260
							24. 602
							  - 333

25.   676
  - 484
	26. 607
	  - 557
		27. 154
		  - 123
			28. 800
			  - 129
				29. 986
				  - 644
					30. 662
					  - 472
						31. 593
						  - 369
							32. 568
							  - 442

33.   446
  - 307
	34. 424
	  - 252
		35. 798
		  - 475
			36. 759
			  - 491
				37. 912
				  - 144
					38. 495
					  - 271
						39. 704
						  - 158
							40. 883
							  - 105

41.   810
  - 718
	42. 887
	  - 593
		43. 439
		  - 322
			44. 296
			  - 198
				45. 289
				  - 108
					46. 323
					  - 310
						47. 832
						  - 670
							48. 641
							  - 488

Find the difference

Complete all the activities (Subtraction)

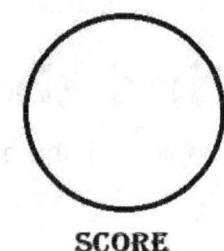

SCORE

Name:.................................   Date:....................................

| 1. | 844
- 796
48 | 2. | 907
- 151
756 | 3. | 109
- 108
1 | 4. | 156
- 146
10 | 5. | 704
- 669
35 | 6. | 887
- 321
566 | 7. | 512
- 454
58 | 8. | 793
- 299
494 |

| 9. | 130
- 110
20 | 10. | 285
- 230
55 | 11. | 871
- 706
165 | 12. | 954
- 342
612 | 13. | 494
- 237
257 | 14. | 402
- 256
146 | 15. | 364
- 348
16 | 16. | 600
- 239
361 |

| 17. | 928
- 918
10 | 18. | 174
- 117
57 | 19. | 598
- 319
279 | 20. | 495
- 160
335 | 21. | 336
- 200
136 | 22. | 204
- 189
15 | 23. | 471
- 260
211 | 24. | 602
- 333
269 |

| 25. | 676
- 484
192 | 26. | 607
- 557
50 | 27. | 154
- 123
31 | 28. | 800
- 129
671 | 29. | 986
- 644
342 | 30. | 662
- 472
190 | 31. | 593
- 369
224 | 32. | 568
- 442
126 |

| 33. | 446
- 307
139 | 34. | 424
- 252
172 | 35. | 798
- 475
323 | 36. | 759
- 491
268 | 37. | 912
- 144
768 | 38. | 495
- 271
224 | 39. | 704
- 158
546 | 40. | 883
- 105
778 |

| 41. | 810
- 718
92 | 42. | 887
- 593
294 | 43. | 439
- 322
117 | 44. | 296
- 198
98 | 45. | 289
- 108
181 | 46. | 323
- 310
13 | 47. | 832
- 670
162 | 48. | 641
- 488
153 |

Name:.................................   Date:....................................

Find the difference

Complete all the activities (Subtraction)

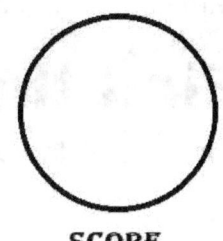

SCORE

1. 912 - 412 --------	2. 521 - 443 --------	3. 165 - 106 --------	4. 719 - 165 --------	5. 940 - 915 --------	6. 997 - 105 --------	7. 984 - 406 --------	8. 432 - 395 --------
9. 768 - 513 --------	10. 884 - 632 --------	11. 703 - 639 --------	12. 419 - 333 --------	13. 441 - 394 --------	14. 334 - 272 --------	15. 359 - 305 --------	16. 800 - 188 --------
17. 507 - 394 --------	18. 576 - 237 --------	19. 664 - 131 --------	20. 941 - 690 --------	21. 264 - 226 --------	22. 722 - 353 --------	23. 691 - 226 --------	24. 563 - 500 --------
25. 704 - 317 --------	26. 277 - 135 --------	27. 914 - 852 --------	28. 142 - 100 --------	29. 127 - 120 --------	30. 492 - 182 --------	31. 844 - 809 --------	32. 834 - 203 --------
33. 271 - 249 --------	34. 597 - 230 --------	35. 834 - 160 --------	36. 336 - 262 --------	37. 714 - 205 --------	38. 545 - 500 --------	39. 282 - 272 --------	40. 166 - 111 --------
41. 412 - 224 --------	42. 157 - 153 --------	43. 636 - 412 --------	44. 297 - 204 --------	45. 707 - 416 --------	46. 331 - 305 --------	47. 657 - 504 --------	48. 886 - 281 --------

Find the difference

Complete all the activities (Subtraction)

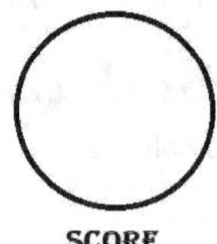

SCORE

1. 912 - 412 500	2. 521 - 443 78	3. 165 - 106 59	4. 719 - 165 554	5. 940 - 915 25	6. 997 - 105 892	7. 984 - 406 578	8. 432 - 395 37
9. 768 - 513 255	10. 884 - 632 252	11. 703 - 639 64	12. 419 - 333 86	13. 441 - 394 47	14. 334 - 272 62	15. 359 - 305 54	16. 800 - 188 612
17. 507 - 394 113	18. 576 - 237 339	19. 664 - 131 533	20. 941 - 690 251	21. 264 - 226 38	22. 722 - 353 369	23. 691 - 226 465	24. 563 - 500 63
25. 704 - 317 387	26. 277 - 135 142	27. 914 - 852 62	28. 142 - 100 42	29. 127 - 120 7	30. 492 - 182 310	31. 844 - 809 35	32. 834 - 203 631
33. 271 - 249 22	34. 597 - 230 367	35. 834 - 160 674	36. 336 - 262 74	37. 714 - 205 509	38. 545 - 500 45	39. 282 - 272 10	40. 166 - 111 55
41. 412 - 224 188	42. 157 - 153 4	43. 636 - 412 224	44. 297 - 204 93	45. 707 - 416 291	46. 331 - 305 26	47. 657 - 504 153	48. 886 - 281 605

Find the difference

Complete all the activities (Subtraction)

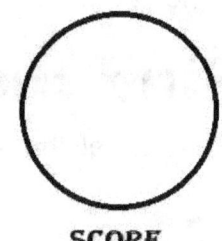

SCORE

1. 959 - 574 --------	2. 295 - 253 --------	3. 693 - 250 --------	4. 143 - 125 --------	5. 186 - 144 --------	6. 562 - 440 --------	7. 485 - 157 --------	8. 826 - 160 --------
9. 698 - 261 --------	10. 875 - 434 --------	11. 899 - 808 --------	12. 589 - 398 --------	13. 688 - 114 --------	14. 627 - 256 --------	15. 842 - 378 --------	16. 161 - 158 --------
17. 367 - 171 --------	18. 949 - 652 --------	19. 757 - 290 --------	20. 556 - 143 --------	21. 413 - 249 --------	22. 849 - 165 --------	23. 161 - 130 --------	24. 190 - 147 --------
25. 741 - 212 --------	26. 300 - 170 --------	27. 876 - 443 --------	28. 407 - 121 --------	29. 106 - 104 --------	30. 217 - 174 --------	31. 699 - 550 --------	32. 787 - 596 --------
33. 497 - 321 --------	34. 494 - 164 --------	35. 283 - 129 --------	36. 416 - 219 --------	37. 785 - 626 --------	38. 175 - 131 --------	39. 147 - 121 --------	40. 586 - 182 --------
41. 931 - 149 --------	42. 624 - 441 --------	43. 236 - 198 --------	44. 185 - 173 --------	45. 931 - 827 --------	46. 679 - 171 --------	47. 243 - 225 --------	48. 485 - 330 --------

Find the difference

Complete all the activities (Subtraction)

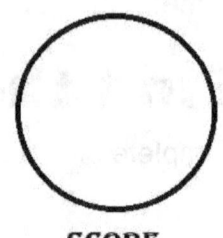

SCORE

1.	2.	3.	4.	5.	6.	7.	8.
959	295	693	143	186	562	485	826
- 574	- 253	- 250	- 125	- 144	- 440	- 157	- 160
385	42	443	18	42	122	328	666

9.	10.	11.	12.	13.	14.	15.	16.
698	875	899	589	688	627	842	161
- 261	- 434	- 808	- 398	- 114	- 256	- 378	- 158
437	441	91	191	574	371	464	3

17.	18.	19.	20.	21.	22.	23.	24.
367	949	757	556	413	849	161	190
- 171	- 652	- 290	- 143	- 249	- 165	- 130	- 147
196	297	467	413	164	684	31	43

25.	26.	27.	28.	29.	30.	31.	32.
741	300	876	407	106	217	699	787
- 212	- 170	- 443	- 121	- 104	- 174	- 550	- 596
529	130	433	286	2	43	149	191

33.	34.	35.	36.	37.	38.	39.	40.
497	494	283	416	785	175	147	586
- 321	- 164	- 129	- 219	- 626	- 131	- 121	- 182
176	330	154	197	159	44	26	404

41.	42.	43.	44.	45.	46.	47.	48.
931	624	236	185	931	679	243	485
- 149	- 441	- 198	- 173	- 827	- 171	- 225	- 330
782	183	38	12	104	508	18	155

Name:............................... Date:...............................

Find the difference

Complete all the activities (Subtraction)

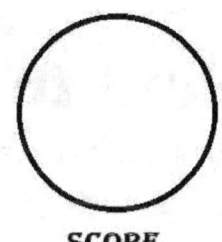

SCORE

1. 738 - 171	2. 680 - 397	3. 141 - 138	4. 390 - 261	5. 840 - 727	6. 629 - 150	7. 423 - 400	8. 115 - 108
--------	--------	--------	--------	--------	--------	--------	--------

9. 161 - 159	10. 739 - 488	11. 251 - 117	12. 613 - 472	13. 515 - 343	14. 315 - 171	15. 561 - 205	16. 295 - 274
--------	--------	--------	--------	--------	--------	--------	--------

17. 406 - 249	18. 733 - 696	19. 830 - 121	20. 211 - 210	21. 762 - 118	22. 843 - 326	23. 215 - 215	24. 545 - 543
--------	--------	--------	--------	--------	--------	--------	--------

25. 872 - 173	26. 150 - 110	27. 244 - 218	28. 485 - 329	29. 586 - 547	30. 145 - 121	31. 727 - 723	32. 139 - 123
--------	--------	--------	--------	--------	--------	--------	--------

33. 666 - 495	34. 779 - 693	35. 301 - 115	36. 763 - 126	37. 852 - 644	38. 883 - 211	39. 359 - 169	40. 636 - 348
--------	--------	--------	--------	--------	--------	--------	--------

41. 181 - 120	42. 523 - 277	43. 256 - 103	44. 756 - 441	45. 153 - 106	46. 459 - 420	47. 980 - 683	48. 361 - 294
--------	--------	--------	--------	--------	--------	--------	--------

Name:.................................   Date:...............................

Find the difference

Complete all the activities (Subtraction)

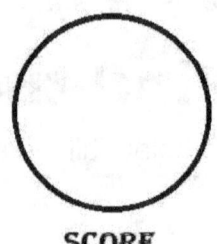

SCORE

1. 738
  - 171
  567

2. 680
  - 397
  283

3. 141
  - 138
  3

4. 390
  - 261
  129

5. 840
  - 727
  113

6. 629
  - 150
  479

7. 423
  - 400
  23

8. 115
  - 108
  7

9. 161
  - 159
  2

10. 739
  - 488
  251

11. 251
  - 117
  134

12. 613
  - 472
  141

13. 515
  - 343
  172

14. 315
  - 171
  144

15. 561
  - 205
  356

16. 295
  - 274
  21

17. 406
  - 249
  157

18. 733
  - 696
  37

19. 830
  - 121
  709

20. 211
  - 210
  1

21. 762
  - 118
  644

22. 843
  - 326
  517

23. 215
  - 215
  0

24. 545
  - 543
  2

25. 872
  - 173
  699

26. 150
  - 110
  40

27. 244
  - 218
  26

28. 485
  - 329
  156

29. 586
  - 547
  39

30. 145
  - 121
  24

31. 727
  - 723
  4

32. 139
  - 123
  16

33. 666
  - 495
  171

34. 779
  - 693
  86

35. 301
  - 115
  186

36. 763
  - 126
  637

37. 852
  - 644
  208

38. 883
  - 211
  672

39. 359
  - 169
  190

40. 636
  - 348
  288

41. 181
  - 120
  61

42. 523
  - 277
  246

43. 256
  - 103
  153

44. 756
  - 441
  315

45. 153
  - 106
  47

46. 459
  - 420
  39

47. 980
  - 683
  297

48. 361
  - 294
  67

Name:................................   Date:...............................

Find the difference

Complete all the activities (Subtraction)

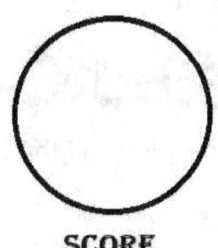

SCORE

1. 983 - 305 --------	2. 707 - 361 --------	3. 319 - 236 --------	4. 794 - 643 --------	5. 420 - 105 --------	6. 426 - 243 --------	7. 954 - 344 --------	8. 377 - 176 --------
9. 391 - 271 --------	10. 434 - 252 --------	11. 570 - 394 --------	12. 479 - 332 --------	13. 971 - 305 --------	14. 832 - 185 --------	15. 708 - 615 --------	16. 154 - 109 --------
17. 489 - 150 --------	18. 243 - 108 --------	19. 632 - 189 --------	20. 297 - 192 --------	21. 523 - 391 --------	22. 156 - 109 --------	23. 689 - 272 --------	24. 612 - 495 --------
25. 820 - 190 --------	26. 610 - 300 --------	27. 362 - 319 --------	28. 297 - 171 --------	29. 828 - 145 --------	30. 514 - 103 --------	31. 826 - 576 --------	32. 196 - 149 --------
33. 959 - 263 --------	34. 104 - 103 --------	35. 214 - 129 --------	36. 714 - 196 --------	37. 430 - 398 --------	38. 984 - 838 --------	39. 606 - 233 --------	40. 961 - 683 --------
41. 699 - 125 --------	42. 384 - 256 --------	43. 351 - 268 --------	44. 893 - 573 --------	45. 917 - 755 --------	46. 321 - 217 --------	47. 735 - 235 --------	48. 694 - 584 --------

Find the difference
Complete all the activities (Subtraction)

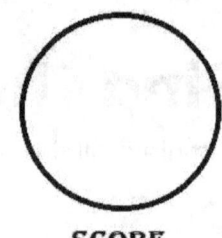

SCORE

1. 983	2. 707	3. 319	4. 794	5. 420	6. 426	7. 954	8. 377
- 305	- 361	- 236	- 643	- 105	- 243	- 344	- 176
678	346	83	151	315	183	610	201

9. 391	10. 434	11. 570	12. 479	13. 971	14. 832	15. 708	16. 154
- 271	- 252	- 394	- 332	- 305	- 185	- 615	- 109
120	182	176	147	666	647	93	45

17. 489	18. 243	19. 632	20. 297	21. 523	22. 156	23. 689	24. 612
- 150	- 108	- 189	- 192	- 391	- 109	- 272	- 495
339	135	443	105	132	47	417	117

25. 820	26. 610	27. 362	28. 297	29. 828	30. 514	31. 826	32. 196
- 190	- 300	- 319	- 171	- 145	- 103	- 576	- 149
630	310	43	126	683	411	250	47

33. 959	34. 104	35. 214	36. 714	37. 430	38. 984	39. 606	40. 961
- 263	- 103	- 129	- 196	- 398	- 838	- 233	- 683
696	1	85	518	32	146	373	278

41. 699	42. 384	43. 351	44. 893	45. 917	46. 321	47. 735	48. 694
- 125	- 256	- 268	- 573	- 755	- 217	- 235	- 584
574	128	83	320	162	104	500	110

Name:................................. Date:.................................

Find the difference

Complete all the activities (Subtraction)

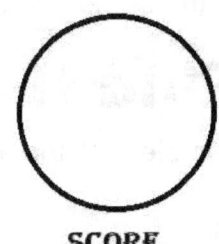

SCORE

1. 622 - 388	2. 282 - 185	3. 359 - 294	4. 681 - 419	5. 497 - 263	6. 123 - 109	7. 867 - 395	8. 815 - 491
9. 194 - 174	10. 407 - 335	11. 406 - 270	12. 308 - 160	13. 377 - 213	14. 921 - 367	15. 105 - 101	16. 505 - 163
17. 318 - 137	18. 329 - 102	19. 155 - 127	20. 450 - 421	21. 456 - 373	22. 199 - 186	23. 677 - 385	24. 769 - 451
25. 105 - 103	26. 482 - 453	27. 883 - 800	28. 526 - 111	29. 389 - 364	30. 241 - 224	31. 851 - 341	32. 588 - 240
33. 964 - 852	34. 231 - 190	35. 217 - 192	36. 536 - 367	37. 455 - 250	38. 433 - 291	39. 501 - 245	40. 680 - 573
41. 697 - 332	42. 387 - 188	43. 756 - 446	44. 356 - 189	45. 725 - 243	46. 719 - 191	47. 402 - 254	48. 174 - 106

Find the difference

Complete all the activities (Subtraction)

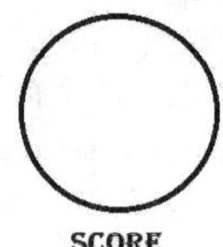

SCORE

1. 622 - 388 234	2. 282 - 185 97	3. 359 - 294 65	4. 681 - 419 262	5. 497 - 263 234	6. 123 - 109 14	7. 867 - 395 472	8. 815 - 491 324
9. 194 - 174 20	10. 407 - 335 72	11. 406 - 270 136	12. 308 - 160 148	13. 377 - 213 164	14. 921 - 367 554	15. 105 - 101 4	16. 505 - 163 342
17. 318 - 137 181	18. 329 - 102 227	19. 155 - 127 28	20. 450 - 421 29	21. 456 - 373 83	22. 199 - 186 13	23. 677 - 385 292	24. 769 - 451 318
25. 105 - 103 2	26. 482 - 453 29	27. 883 - 800 83	28. 526 - 111 415	29. 389 - 364 25	30. 241 - 224 17	31. 851 - 341 510	32. 588 - 240 348
33. 964 - 852 112	34. 231 - 190 41	35. 217 - 192 25	36. 536 - 367 169	37. 455 - 250 205	38. 433 - 291 142	39. 501 - 245 256	40. 680 - 573 107
41. 697 - 332 365	42. 387 - 188 199	43. 756 - 446 310	44. 356 - 189 167	45. 725 - 243 482	46. 719 - 191 528	47. 402 - 254 148	48. 174 - 106 68

Name:..............................   Date:..................................

Find the difference

Complete all the activities (Subtraction)

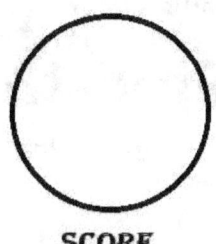

SCORE

1.   631
   - 627
   -------- 	
	2. 342
	   - 267
	   -------- 	
		3. 880
		   - 387
		   -------- 	
			4. 262
			   - 184
			   -------- 	
				5. 790
				   - 541
				   -------- 	
					6. 953
					   - 912
					   -------- 	
						7. 897
						   - 219
						   -------- 	
							8. 949
							   - 389
							   -------- 	

9.   204
   - 107
   -------- 	
	10. 784
	   - 159
	   -------- 	
		11. 866
		   - 159
		   -------- 	
			12. 496
			   - 417
			   -------- 	
				13. 768
				   - 738
				   -------- 	
					14. 487
					   - 166
					   -------- 	
						15. 145
						   - 102
						   -------- 	
							16. 655
							   - 351
							   -------- 	

17.   553
   - 253
   -------- 	
	18. 117
	   - 111
	   -------- 	
		19. 609
		   - 589
		   -------- 	
			20. 742
			   - 730
			   -------- 	
				21. 577
				   - 201
				   -------- 	
					22. 871
					   - 124
					   -------- 	
						23. 677
						   - 383
						   -------- 	
							24. 907
							   - 287
							   -------- 	

25.   382
   - 343
   -------- 	
	26. 571
	   - 543
	   -------- 	
		27. 144
		   - 115
		   -------- 	
			28. 662
			   - 389
			   -------- 	
				29. 597
				   - 333
				   -------- 	
					30. 953
					   - 878
					   -------- 	
						31. 943
						   - 108
						   -------- 	
							32. 519
							   - 399
							   -------- 	

33.   120
   - 110
   -------- 	
	34. 503
	   - 321
	   -------- 	
		35. 217
		   - 135
		   -------- 	
			36. 796
			   - 208
			   -------- 	
				37. 418
				   - 348
				   -------- 	
					38. 599
					   - 420
					   -------- 	
						39. 297
						   - 181
						   -------- 	
							40. 244
							   - 136
							   -------- 	

41.   152
   - 110
   -------- 	
	42. 871
	   - 246
	   -------- 	
		43. 577
		   - 423
		   -------- 	
			44. 462
			   - 162
			   -------- 	
				45. 741
				   - 424
				   -------- 	
					46. 187
					   - 109
					   -------- 	
						47. 111
						   - 109
						   -------- 	
							48. 139
							   - 120
							   -------- 	

Name:............................ Date:..............................

Find the difference

Complete all the activities (Subtraction)

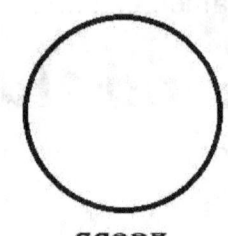

SCORE

1. 631 - 627 4	2. 342 - 267 75	3. 880 - 387 493	4. 262 - 184 78	5. 790 - 541 249	6. 953 - 912 41	7. 897 - 219 678	8. 949 - 389 560
9. 204 - 107 97	10. 784 - 159 625	11. 866 - 159 707	12. 496 - 417 79	13. 768 - 738 30	14. 487 - 166 321	15. 145 - 102 43	16. 655 - 351 304
17. 553 - 253 300	18. 117 - 111 6	19. 609 - 589 20	20. 742 - 730 12	21. 577 - 201 376	22. 871 - 124 747	23. 677 - 383 294	24. 907 - 287 620
25. 382 - 343 39	26. 571 - 543 28	27. 144 - 115 29	28. 662 - 389 273	29. 597 - 333 264	30. 953 - 878 75	31. 943 - 108 835	32. 519 - 399 120
33. 120 - 110 10	34. 503 - 321 182	35. 217 - 135 82	36. 796 - 208 588	37. 418 - 348 70	38. 599 - 420 179	39. 297 - 181 116	40. 244 - 136 108
41. 152 - 110 42	42. 871 - 246 625	43. 577 - 423 154	44. 462 - 162 300	45. 741 - 424 317	46. 187 - 109 78	47. 111 - 109 2	48. 139 - 120 19

Name:................................   Date:................................

Find the difference

Complete all the activities (Subtraction)

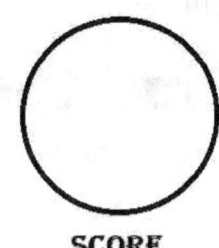

SCORE

1. 769	2. 108	3. 139	4. 191	5. 178	6. 627	7. 799	8. 365
- 196	- 101	- 114	- 102	- 129	- 242	- 788	- 138
--------	--------	--------	--------	--------	--------	--------	--------

9. 377	10. 868	11. 237	12. 264	13. 184	14. 851	15. 279	16. 229
- 293	- 667	- 138	- 157	- 123	- 570	- 221	- 171
--------	--------	--------	--------	--------	--------	--------	--------

17. 763	18. 906	19. 461	20. 287	21. 188	22. 192	23. 182	24. 453
- 122	- 161	- 105	- 232	- 174	- 115	- 111	- 357
--------	--------	--------	--------	--------	--------	--------	--------

25. 289	26. 810	27. 953	28. 433	29. 965	30. 146	31. 562	32. 836
- 280	- 324	- 360	- 256	- 674	- 135	- 320	- 403
--------	--------	--------	--------	--------	--------	--------	--------

33. 303	34. 579	35. 179	36. 767	37. 539	38. 676	39. 258	40. 328
- 107	- 457	- 117	- 296	- 164	- 280	- 103	- 319
--------	--------	--------	--------	--------	--------	--------	--------

41. 895	42. 474	43. 897	44. 641	45. 696	46. 202	47. 966	48. 275
- 185	- 192	- 760	- 334	- 356	- 175	- 483	- 209
--------	--------	--------	--------	--------	--------	--------	--------

Name:..................................   Date:..................................

Find the difference

Complete all the activities (Subtraction)

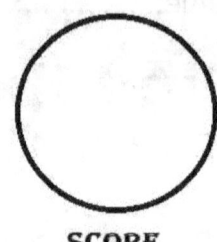

SCORE

1. 769	2. 108	3. 139	4. 191	5. 178	6. 627	7. 799	8. 365
- 196	- 101	- 114	- 102	- 129	- 242	- 788	- 138
573	7	25	89	49	385	11	227

9. 377	10. 868	11. 237	12. 264	13. 184	14. 851	15. 279	16. 229
- 293	- 667	- 138	- 157	- 123	- 570	- 221	- 171
84	201	99	107	61	281	58	58

17. 763	18. 906	19. 461	20. 287	21. 188	22. 192	23. 182	24. 453
- 122	- 161	- 105	- 232	- 174	- 115	- 111	- 357
641	745	356	55	14	77	71	96

25. 289	26. 810	27. 953	28. 433	29. 965	30. 146	31. 562	32. 836
- 280	- 324	- 360	- 256	- 674	- 135	- 320	- 403
9	486	593	177	291	11	242	433

33. 303	34. 579	35. 179	36. 767	37. 539	38. 676	39. 258	40. 328
- 107	- 457	- 117	- 296	- 164	- 280	- 103	- 319
196	122	62	471	375	396	155	9

41. 895	42. 474	43. 897	44. 641	45. 696	46. 202	47. 966	48. 275
- 185	- 192	- 760	- 334	- 356	- 175	- 483	- 209
710	282	137	307	340	27	483	66

Find the difference

Complete all the activities (Subtraction)

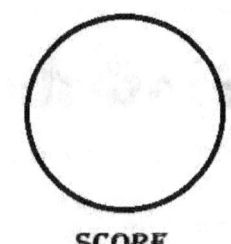

SCORE

Name:.................................. Date:..................................

1. 118 - 112 --------	2. 942 - 147 --------	3. 640 - 450 --------	4. 676 - 670 --------	5. 291 - 195 --------

1. 118 − 112 ———
2. 942 − 147 ———
3. 640 − 450 ———
4. 676 − 670 ———
5. 291 − 195 ———
6. 843 − 697 ———
7. 697 − 134 ———
8. 362 − 347 ———

9. 469 − 134 ———
10. 893 − 423 ———
11. 153 − 130 ———
12. 572 − 288 ———
13. 490 − 139 ———
14. 174 − 149 ———
15. 196 − 100 ———
16. 398 − 114 ———

17. 955 − 115 ———
18. 363 − 227 ———
19. 175 − 115 ———
20. 423 − 176 ———
21. 518 − 278 ———
22. 130 − 112 ———
23. 102 − 102 ———
24. 791 − 518 ———

25. 224 − 121 ———
26. 393 − 355 ———
27. 604 − 496 ———
28. 607 − 326 ———
29. 250 − 181 ———
30. 659 − 380 ———
31. 609 − 601 ———
32. 895 − 265 ———

33. 248 − 215 ———
34. 866 − 595 ———
35. 787 − 655 ———
36. 872 − 265 ———
37. 758 − 524 ———
38. 947 − 637 ———
39. 586 − 165 ———
40. 456 − 326 ———

41. 397 − 241 ———
42. 567 − 536 ———
43. 710 − 380 ———
44. 537 − 172 ———
45. 508 − 445 ———
46. 386 − 256 ———
47. 617 − 544 ———
48. 703 − 391 ———

Name:.................................   Date:................................

Find the difference
Complete all the activities (Subtraction)

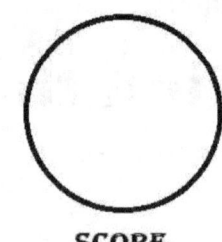

SCORE

1. 118 - 112 6	2. 942 - 147 795	3. 640 - 450 190	4. 676 - 670 6	5. 291 - 195 96	6. 843 - 697 146	7. 697 - 134 563	8. 362 - 347 15
9. 469 - 134 335	10. 893 - 423 470	11. 153 - 130 23	12. 572 - 288 284	13. 490 - 139 351	14. 174 - 149 25	15. 196 - 100 96	16. 398 - 114 284
17. 955 - 115 840	18. 363 - 227 136	19. 175 - 115 60	20. 423 - 176 247	21. 518 - 278 240	22. 130 - 112 18	23. 102 - 102 0	24. 791 - 518 273
25. 224 - 121 103	26. 393 - 355 38	27. 604 - 496 108	28. 607 - 326 281	29. 250 - 181 69	30. 659 - 380 279	31. 609 - 601 8	32. 895 - 265 630
33. 248 - 215 33	34. 866 - 595 271	35. 787 - 655 132	36. 872 - 265 607	37. 758 - 524 234	38. 947 - 637 310	39. 586 - 165 421	40. 456 - 326 130
41. 397 - 241 156	42. 567 - 536 31	43. 710 - 380 330	44. 537 - 172 365	45. 508 - 445 63	46. 386 - 256 130	47. 617 - 544 73	48. 703 - 391 312

Name:...................................   Date:...................................

Find the difference
Complete all the activities (Subtraction)

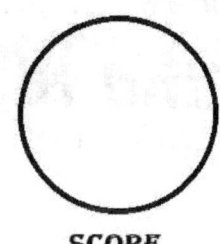

SCORE

1. 587 - 502 --------	2. 303 - 145 --------	3. 951 - 205 --------	4. 856 - 350 --------	5. 381 - 132 --------	6. 225 - 214 --------	7. 859 - 806 --------	8. 914 - 126 --------
9. 229 - 227 --------	10. 253 - 114 --------	11. 517 - 496 --------	12. 576 - 436 --------	13. 363 - 281 --------	14. 272 - 258 --------	15. 540 - 239 --------	16. 674 - 602 --------
17. 976 - 503 --------	18. 183 - 159 --------	19. 917 - 242 --------	20. 720 - 229 --------	21. 671 - 208 --------	22. 659 - 416 --------	23. 388 - 265 --------	24. 388 - 251 --------
25. 946 - 342 --------	26. 460 - 312 --------	27. 718 - 459 --------	28. 116 - 109 --------	29. 366 - 239 --------	30. 936 - 317 --------	31. 937 - 872 --------	32. 856 - 854 --------
33. 527 - 213 --------	34. 946 - 425 --------	35. 853 - 243 --------	36. 211 - 208 --------	37. 366 - 356 --------	38. 509 - 212 --------	39. 395 - 385 --------	40. 446 - 350 --------
41. 429 - 320 --------	42. 979 - 279 --------	43. 922 - 140 --------	44. 267 - 166 --------	45. 497 - 298 --------	46. 478 - 247 --------	47. 319 - 194 --------	48. 348 - 277 --------

Name:.................................. Date:..................................

Find the difference
Complete all the activities (Subtraction)

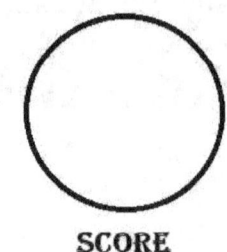

SCORE

1. 587	2. 303	3. 951	4. 856	5. 381	6. 225	7. 859	8. 914
- 502	- 145	- 205	- 350	- 132	- 214	- 806	- 126
85	158	746	506	249	11	53	788

9. 229	10. 253	11. 517	12. 576	13. 363	14. 272	15. 540	16. 674
- 227	- 114	- 496	- 436	- 281	- 258	- 239	- 602
2	139	21	140	82	14	301	72

17. 976	18. 183	19. 917	20. 720	21. 671	22. 659	23. 388	24. 388
- 503	- 159	- 242	- 229	- 208	- 416	- 265	- 251
473	24	675	491	463	243	123	137

25. 946	26. 460	27. 718	28. 116	29. 366	30. 936	31. 937	32. 856
- 342	- 312	- 459	- 109	- 239	- 317	- 872	- 854
604	148	259	7	127	619	65	2

33. 527	34. 946	35. 853	36. 211	37. 366	38. 509	39. 395	40. 446
- 213	- 425	- 243	- 208	- 356	- 212	- 385	- 350
314	521	610	3	10	297	10	96

41. 429	42. 979	43. 922	44. 267	45. 497	46. 478	47. 319	48. 348
- 320	- 279	- 140	- 166	- 298	- 247	- 194	- 277
109	700	782	101	199	231	125	71

Name:............................ Date:............................

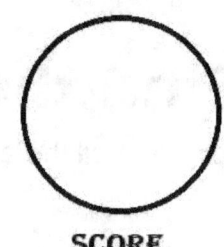

SCORE

Find the difference

Complete all the activities (Subtraction)

1. 413 - 295	2. 181 - 104	3. 667 - 631	4. 717 - 563	5. 246 - 163	6. 364 - 291	7. 214 - 185	8. 338 - 179
9. 222 - 125	10. 902 - 312	11. 949 - 641	12. 175 - 132	13. 171 - 139	14. 947 - 387	15. 792 - 609	16. 753 - 665
17. 717 - 558	18. 932 - 686	19. 886 - 360	20. 182 - 135	21. 982 - 301	22. 137 - 118	23. 779 - 413	24. 340 - 268
25. 311 - 119	26. 327 - 108	27. 584 - 106	28. 791 - 592	29. 586 - 423	30. 976 - 173	31. 994 - 411	32. 982 - 476
33. 733 - 371	34. 135 - 113	35. 835 - 512	36. 511 - 431	37. 205 - 149	38. 355 - 283	39. 635 - 420	40. 702 - 472
41. 352 - 162	42. 193 - 163	43. 535 - 400	44. 394 - 298	45. 975 - 288	46. 247 - 131	47. 420 - 103	48. 492 - 348

Name:................................ Date:................................

Find the difference

Complete all the activities (Subtraction)

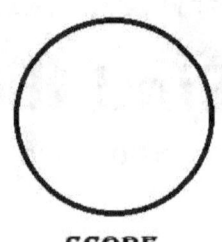

SCORE

1. 413 - 295 = 118	2. 181 - 104 = 77	3. 667 - 631 = 36	4. 717 - 563 = 154	5. 246 - 163 = 83	6. 364 - 291 = 73	7. 214 - 185 = 29	8. 338 - 179 = 159
9. 222 - 125 = 97	10. 902 - 312 = 590	11. 949 - 641 = 308	12. 175 - 132 = 43	13. 171 - 139 = 32	14. 947 - 387 = 560	15. 792 - 609 = 183	16. 753 - 665 = 88
17. 717 - 558 = 159	18. 932 - 686 = 246	19. 886 - 360 = 526	20. 182 - 135 = 47	21. 982 - 301 = 681	22. 137 - 118 = 19	23. 779 - 413 = 366	24. 340 - 268 = 72
25. 311 - 119 = 192	26. 327 - 108 = 219	27. 584 - 106 = 478	28. 791 - 592 = 199	29. 586 - 423 = 163	30. 976 - 173 = 803	31. 994 - 411 = 583	32. 982 - 476 = 506
33. 733 - 371 = 362	34. 135 - 113 = 22	35. 835 - 512 = 323	36. 511 - 431 = 80	37. 205 - 149 = 56	38. 355 - 283 = 72	39. 635 - 420 = 215	40. 702 - 472 = 230
41. 352 - 162 = 190	42. 193 - 163 = 30	43. 535 - 400 = 135	44. 394 - 298 = 96	45. 975 - 288 = 687	46. 247 - 131 = 116	47. 420 - 103 = 317	48. 492 - 348 = 144